International Laser Technology and Optics Symposium (ILATOS 2017)

Journal of Physics: Conference Series Volume 1027

Johor Bahru, Malaysia
26 – 28 September 2017

ISBN: 978-1-5108-6345-3
ISSN: 1742-6588

Printed from e-media with permission by:

Curran Associates, Inc.
57 Morehouse Lane
Red Hook, NY 12571

CURRAN ASSOCIATES INC. proceedings.com

Some format issues inherent in the e-media version may also appear in this print version.

Printed by Curran Associates, Inc. (2018)

Institute of Physics
Dirac House, Temple Back
Bristol BS1 6BE UK

Phone: 44 1 17 929 7481
Fax: 44 1 17 920 0979

techtracking@iop.org

Additional copies of this publication are available from:

Curran Associates, Inc.
57 Morehouse Lane
Red Hook, NY 12571 USA
Phone: 845-758-0400
Fax: 845-758-2633
Email: curran@proceedings.com
Web: www.proceedings.com

Table of contents

Volume 1027

International Laser Technology and Optics Symposium (ILATOS 2017)

26–28 September 2017, Johor Bahru, Malaysia

Accepted papers received: 8 May 2018 Published online: 22 May 2018

Preface

Papers

IOP Conf. Series: Journal of Physics: Conf. Series **1027** (2018) 011001 doi:10.1088/1742-6596/1027/1/011001

ILATOS 2017

Dear Readers,

The annual International Laser Technology and Optics Symposium 2017 (ILATOS 2017) is an international forum for researchers, academicians, engineers and industry players from various fields of laser & electro-optics with different experiences and backgrounds. ILATOS 2017 creates opportunities to present research results, discuss current issues and exchange views that may create mutual interest.

ILATOS 2017 is a bridge to transfer research & development output to the applications which match the industrial requirements as well as user's expectations. ILATOS 2017 offers a comprehensive view of theoretical and experimental laser research and applications. Articles cover every aspect of modern laser physics and quantum electronics, emphasizing physical effects in various media (solid, gaseous, and liquid) leading to the generation of laser radiation; peculiarities of propagation of laser radiation, problems involving impact of laser radiation on various substances and the emerging physical effects, including coherent ones; the applied use of lasers and laser spectroscopy; the processing and storage of information.

The 2017 edition of ILATOS was held from the 26th to 28th of September 2017 in the city of Johor Bahru, at the Pulai Spring Resort, Malaysia.

IOP Conf. Series: Journal of Physics: Conf. Series **1027** (2018) 011002 doi:10.1088/1742-6596/1027/1/011002

Peer review statement

All papers published in this volume of *Journal of Physics: Conference Series* have been peer reviewed through processes administered by the proceedings Editors. Reviews were conducted by expert referees to the professional and scientific standards expected of a proceedings journal published by IOP Publishing.

IOP Conf. Series: Journal of Physics: Conf. Series **1027** (2018) 012001 doi:10.1088/1742-6596/1027/1/012001

Knife edges technique for focused beam profiling of the photothermal imaging system

Azura Amran and Nor Kamilah Saat*

Department of Physics, Faculty of Science, Universiti Putra Malaysia, Malaysia

Corresponding author: kamilah@upm.edu.my

Abstract. In this paper, we demonstrate knife edge techniques for measuring the focused beam profile of the photothermal imaging system. This quick and simple method of profiling the focused beam from Helium-Neon (He-Ne) laser is presented and detected by the PVDF thin film as a sensor. The study of focused beam profile is crucial so as to provide information of the focused beam such as beam resolution, depth of focus and the level of astigmatism of the beam. In knife edge technique, the blade moves across the beam in x and y-direction at varying axial position.

1. Introduction

Performing of beam profile plays the main role in characterize of laser beam because the beam profile provides information about beam resolution, depth of focus as well as the level of astigmatism of the focused beam particularly in an imaging system. In general, there are two types of techniques in characterizing the beam profile, i.e. camera based technique and scanning aperture technique. In camera based technique allows the beam profile to be measured directly in two-dimensional [1]. This technique can be done in real-time, and it only suitable for the visible up to near-infrared spectral range. It also has low saturation and damage threshold levels, which often requires considerable attenuation. Camera based systems work with larger beam and either pulsed or CW laser.

For scanning aperture technique, the light passing through is measured by a detector and correlated with the position of the aperture as it crosses the beam. Scanning aperture techniques includes the pinhole technique [2], the slit technique [3] and knife-edge technique [4]. In pinhole profiles use small pinhole as the aperture and plot the transmitted power against the position. The determination of the profile is dictated by the size of the aperture. In slit profile utilize long and narrow aperture which encompassed the full width of the beam in a direction perpendicular to the travel of the slit. However, in knife-edge technique, the aperture using sharp and straight edge like knife-edge or blade is passed through the beam. The opaque side the knife blocks out a part of the beam, while the power of the transmitted part of the beam is measured by a detector. The transmitted power for different positions of the knife edge is obtained by scanning the knife edge transversely across the beam along the orthogonal direction to the edge [4]. The knife edge technique is a beam profiling method to determine the beam width of focus beam by assuming the incoming beam is having a Gaussian beam profile [5]. The knife-edge technique is the most popular, cheap and quick technique. It is most useful with very small beams, CW laser and high power laser. However, the drawback of this technique has been the increase of error since to differentiation the step response function (SRF) [6] which will amplify the noise. By fitting Gaussian equation into SRF and extract the beam widths of the beam focus. In this section, the step of knife-edge technique for characterizing the focused beam profile is described.

2. Experimental

The schematic diagram of knife-edge technique experimental is shown in figure 1. The beam of 150mW He-Ne laser modulated by variable frequency optical chopper (SR540) is impinged onto the PVDF film supported by 1.2 cm thick Perspex backing. The output from the pyroelectric detector is fed into a lock-in amplifier (SR530) for signal analysis. Frequency optical chopper is to give pulse laser at 5 Hz frequency modulation.

For the setting up of knife edge technique, a blade is used to measure beam's diameter. The blade was mounted on top of the motorized translation stage. There will give two types of translation, which included longitudinal displacement (z-direction) and the transversal movement (x and y direction). In transversal stages, the translation was mounted to be perpendicular. Besides, the knife edge pair was mounted on the transversal stage and they should align correctly with the ground.

To study the beam profile of the focused beam, a radiation–blocking plate was mounted on a computer-controlled two-dimensional translation stage. This plate was moved across the beam in x (horizontal) and y (vertical) orientation at different position along the z-axis beam using 0.5 mm step size. The step responses for both orientations were recorded, allowing step response function (SRF) to be plotted.

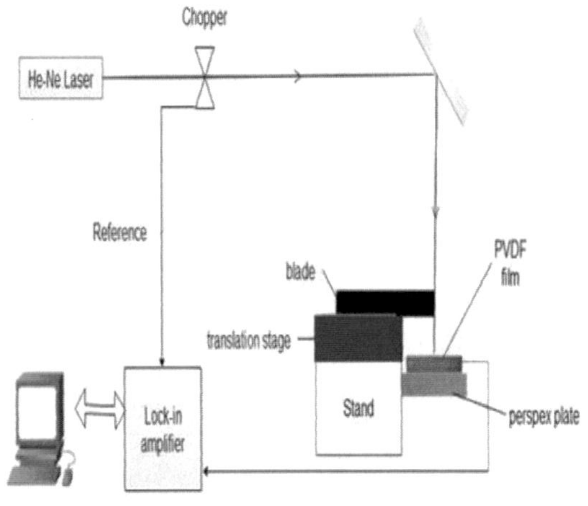

Figure 1. Schematic diagram of knife-edge experimental setup

3. Results and discussion

The trend of intensity that obtained against the x position and y position are still in the Gaussian pattern which assumed that it is the Gaussian beam profile for this research. In order to obtain the beam width of horizontal and vertical orientation, the expression is given by:

$$p = \int_{-\infty}^{\infty}\int_{-\infty}^{\infty} I(x,y)\,dx\,dy = \frac{p_o}{2}\,erfc\left(\frac{a\sqrt{2}}{w}\right) \quad (1)$$

Where I(x,y) is the intensity level at x and y position of the beam that perpendicular to z axis which is the direction of propagation $erfc(a\sqrt{2}/w)$ is an integral of Gaussian.

The fraction p/p_0 rely on the term $(a\sqrt{2}/w)$ where a and w indicate the position of the knife edge and the beam width, respectively. The SRF for the both orientations, show that the fitting for

IOP Conf. Series: Journal of Physics: Conf. Series **1027** (2018) 012001 doi:10.1088/1742-6596/1027/1/012001

transmitted power against x-axis labelled $x - x_0/w$, where x_0 the beam center and x is the distance from the beam center. The gradient of the curve p/p_o is related to the beam width, where the shallower the gradient of p/p_o is the larger the beam width [6].

Figure 2 shows the SRF at different position along the beam axis for both horizontal and vertical orientation, at a frequency of 5 Hz. The SRF at position 5 cm, measured from a beam focus shows the maximum gradient in both horizontal and vertical orientation representing the minimum beam width of the beam focus.

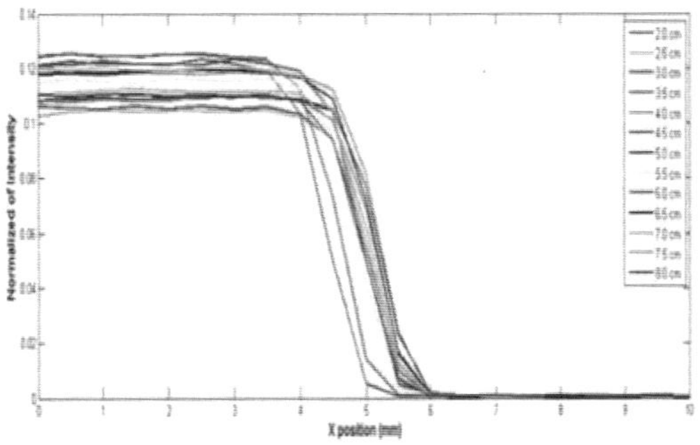

(a)

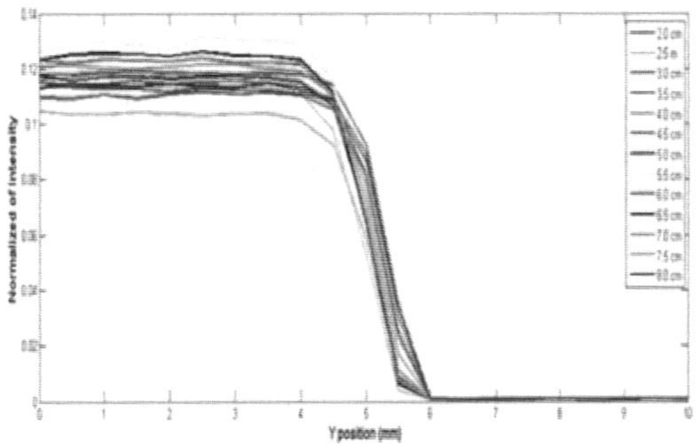

(b)

Figure 2: Step response function (SRF) of (a) horizontal and (b) vertical orientations.

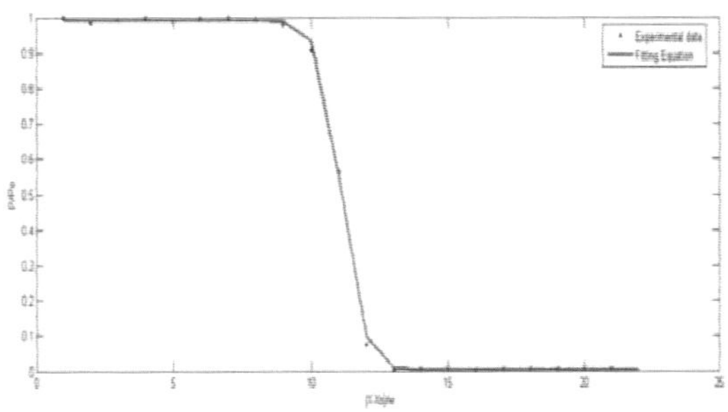

Figure 3: Example of fitting equation to the experimental data to find the beam width in the vertical (y) direction, at a beam axis of 5 cm, measured from the beam focus.

Figure 4 shows the beam width deduced from the SRF in (a) horizontal and (b) vertical directions. The minimum beam width for both orientations was 0.7386 mm in the horizontal and 0.6871 mm in the vertical. The astigmatic difference is the difference between minimum beam waist in the horizontal and vertical orientation is 0.0515 mm. This astigmatism effect was expected, probably due to the beam is not propagated through the center of the mirror.

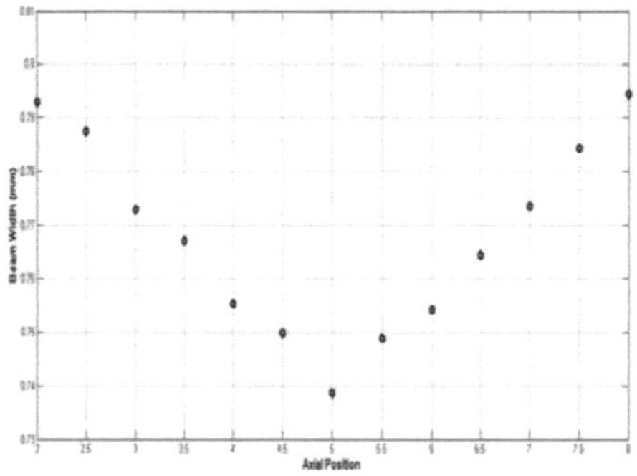

(a)

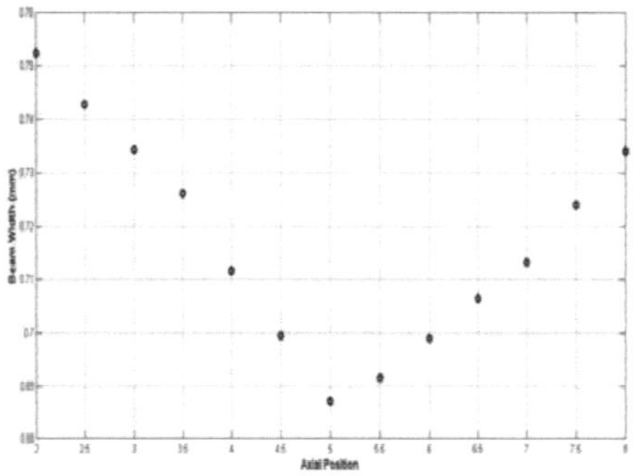

(b)

Figure 4: Beam width w obtained from Gaussian fits to the experimental data for (a) horizontal and (b) vertical orientation

4. Conclusion

In conclusion, the size of the beam focus of the imaging system was characterized along the beam axis using the knife edge technique. The beam width of the focused beam in horizontal and vertical were 0.7386 mm and 0.6871 mm respectively, with small astigmatism effect.

Acknowledgements
We acknowledge support from the Universiti Putra Malaysia and the Government of Malaysia.

References

[1] Siegman A E 1998 How to (Maybe) Measure Laser Beam Quality *DPSS (Diode Pumped Solid State) Lasers: Applications and Issues*

[2] Boley C D, Estabrook K G, Auerbach J M, Feit M D and Rubenchik A M 1999 Third International Conference on *Solid State Lasers for Application to Inertial Confinement Fusion*

[3] Chapple P B 1994 *Optical Engineering* **33** 2461
[4] González-Cardel M, Arguijo P and Díaz-Uribe R 2013 *Applied Optics* **52** 3849
[5] Khosrofian J M and Garetz B A 1983 *Applied Optics* **22** 3406
[6] Saat N K, Dean P, Khanna S P, Salih M, Linfield E H and Davies A G 2016 *Jurnal Teknologi* **78**
[7] Araújo M A D, Silva R, Lima E D, Pereira D P and Oliveira P C D 2009 *Applied Optics* **48** 393

Optical and structure characterization of cinnamon nanoparticles synthesized by pulse laser ablation in liquid (PLAL)

Ali Aqeel Salim[1], Noriah Bidin*[1], Hazri Bakhtiar[1], Sib Krishna Ghoshal[2], Mohammed Al Azawi[1], Ganesan Krishnan[1]

[1]Laser Center, Ibnu Sina Institute for Scientific and Industrial Research (ISI-SIR)
[2]Advanced Optical Materials Research Group, Physics Department, Faculty of Science,
[1,2]Universiti Teknologi Malaysia, 81310 Johor Bahru, Johor, Malaysia

Corresponding author: noriah@utm.my

Abstract. Organic nanoparticles development is under exploration due to its beneficial applications in nanobiomedical and research interests. PLAL technique of Q-switched 1064-Nd: YAG (10 ns pulse duration, repetition rate 1 Hz and laser energy 20-100 mJ) has inherent advantages and rapid growth of nanoparticles when compared to conventional methods because of the controlled fabricated nanoparticles, stability, and purity. Cinnamon sticks as a target are immersed in 5 ml ethanol medium and irradiated by a laser beam for the growth process. The morphology, optical characteristic, and bonding structure of cinnamon nanoparticles (CNPs) are determined and evaluated by transmission electron microscope (TEM), UV-Visible spectroscopy and Fourier transform infrared spectroscopy (FTIR). Spherical, homogenous and high crystallinity CNPs was revealed within the particle size range of 2 - 28 nm. The absorption band was found in the ultraviolent region around 259 nm and 319 nm. The present of FTIR spectra confirmed that the nanoparticles were covered by plant secondary metabolites. The experimental findings revealed that the synthesize CNPs in ethanol has a potential for nanomedicine applications.

1. Introduction

Nanotechnology is a multidisciplinary field and fast grown, especially in biotechnology [1]. the advanced synthesized and analytical tools of nanoscience in the chemical, physical, biphotonic, biological and electronic can be controlled at the nano-scale due to their distinctive structures and unique properties form [2]. Indeed, nanomaterials are under investigation due to wide-scale usages in the nanobiomedicine, biphotonic applications and research interest in nanoscience; of great systematic interest as they bridge the gap between bulk materials and molecular structures.Other paragraphs are indented (BodytextIndented style).

Recently, there is a growing interest in the organic material of systems at the nanoscale because lots of materials have diverse unique properties when compared to bulk counterpart. Therefore, intensive studies using organic nanoparticles suspended in surfactant solutions have been assumed because of their size-dependent, morphology and characteristic properties [3]. Earlier, several development techniques are applied for the synthesis of nanoparticles having diverse morphologies such as nanowire, nanoflower, nanoprolate, nanotube, nanospheres, nanorods, etc. [4,5]. These approaches techniques are employed for multidisciplinary research area synthesis of nanoparticles including laser ablation, ball milling, precipitation, chemical etching, sputtering, phase transport, sol-gel, vapor deposition, molecular condensation, enzymes and microorganism [5,6]. However, pulse laser ablation in liquid (PLAL) became recently widespread methods to the development of organic nanoparticles and overcome on many other techniques due to its interesting attributes such as simplicity, growth controlled, variability, economic and free of impurities. It has a promising advantage to control the size, shape, and crystallinity of the nanoparticles by changing the laser parameters such as laser fluence, repletion rate, wavelength, pulse number and solvent [6,7].

Natural plant parts appeared as a prospective contender for a broad range of biomedicine applications due to their biocompatibility, variability low cultivation cost, effortlessness, abundance, short production, these are favorited over metal nanoparticles [8]. Therefore, cinnamon parts are attentionally attractive the researchers because of its wide range of beneficial usages and considering as one of the traditional herbal medicine in Asia due to the presence of the natural major compound in cinnamon such as polyphenols and cinnamaldehyde [9].

In this novel report, we produced CNPs immersed in ethanol liquid medium accompanying by different laser ablation energy of 20 mJ up to 120 mJ using PLAL technique. The effect of various laser ablation energy on morphology, absorption properties and chemical bonding structural of CNPs was determined and observed by TEM, UV–Vis and FTIR measurements.

2. Methodology

2.1. Experiment Materials and Sample Preparation

Cinnamon sticks with dimension (80 mm × 20 mm × 2 mm) were purchased from the local supermarket (Aeon, Kuala Lumpur, Malaysia). Analytical grade ethanol (C2H5OH, 96% purity from Sigma Aldrich) was used as liquid media to grow the CNPs. Cinnamon sticks were chopped at a dimension of (20 × 10 × 3) mm and cleaned using ultrasonic bath with acetone solvent, thereafter, each small cinnamon stick washed with purified water to eliminate the presence of any contaminants.

2.2. Synthesis CNPs

Cinnamon nanoparticles grown in ethanol liquid medium was synthesized using PLAL technique. A Q-switched 1064-Nd: YAG laser (10 ns pulse duration, repetition rate 1 Hz and varying laser energy 20-100 mJ) was applied. First, the cinnamon stick as a target material was immersed at the 27 cm^3 cubic container bottom with 5 ml liquid ethanol as growth solution. the schematic diagram of the experimental setup of CNPs displayed in the Figure 1. The laser irradiation was directly focused on the surface of the target through a lens of focal length 80 mm. The distance between the laser lens and cinnamon surface target was positioned at 17 mm because of the refraction of the light phenomenon as reported earlier [10].

IOP Conf. Series: Journal of Physics: Conf. Series **1027** (2018) 012002 doi:10.1088/1742-6596/1027/1/012002

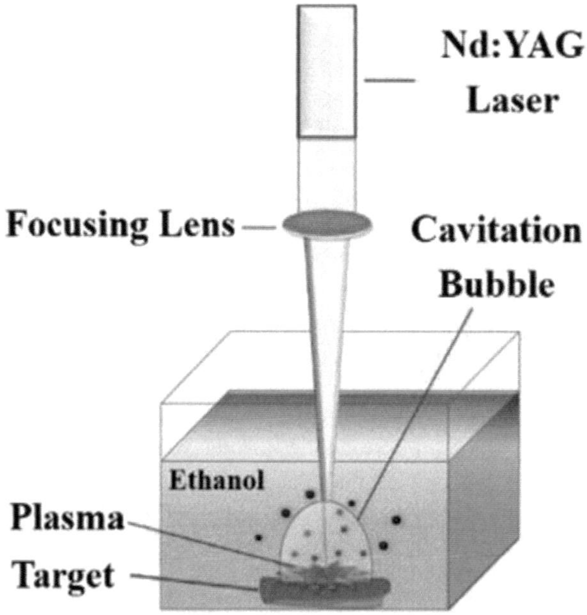

Figure 1. Schematic diagram of the experimental setup for synthesizing CNPs in ethanol

2.3. Characterizations of CNPs

The morphology (size and shape) and fast Fourier transform (FFT) pattern of CNPs were examined via Biological Transmission Electron Microscope (model, BIO-TEM from Hitachi HT7700, Universiti Teknologi Malaysia (UTM)). Absorption spectra of CNPs were measured using a UV-Vis spectrophotometer (PerkinElmer Lambda 25 Spectrometer, UTM). FTIR absorption spectra of the CNPs samples in the wave number range of 500-4000 cm-1 was recorded on a PerkinElmer Frontier™ Spectrometer, UTM using KBr pellets techniques. In this process, a small drop (□0.5 ml) of the sample was poured on one of the KBr pellet surfaces using a Pasteur pipette and the second pellet was placed on the top of the drop with a quarter turn.

3. Results and Discussion

3.1. Morphology of CNPs

Figure 2(a) and (b) show the TEM image, FFT pattern and size distribution of CNPs grown at the optimum laser ablation energy of 80 mJ respectively. The TEM image in Figure 2(a) shows the morphology of the CNPs which was found to be predominantly spherical in shape, homogenous, and small particle diameter. This morphology of CNPs was attributed to the generated atoms that were trapped by the existing nuclei during the diffusion-controlled growth process. The inset in Figure 2(a) depicts the crystallinity phase of CNP. The corresponding (Figure 2 (a)) CNPs size distribution is revealed in Figure 2(b) where the average size was ascertained to be ≈ 9.14 nm.

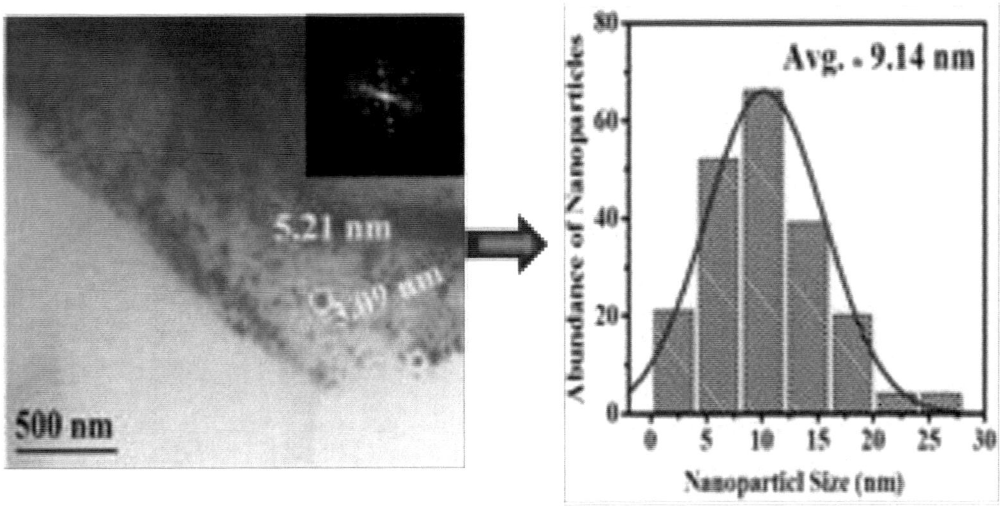

Figure 2. Cinnamon nanoparticles synthesized with optimum laser ablation energy of 80 mJ (a) TEM morphology (Inset: the FFT pattern) (b) size distribution of CNPs corresponding to (a).

3.2. UV-Vis Absorption Bands

Figure 3 illustrates the absorption bands spectra of all prepared samples as a function of laser ablation energy. The CNPs containing solution color and the intensities as well as absorption peak positions was changed which indicating the difference growth in the CNPs morphology (size and shapes) and crystallinity. This observation agreed well with the findings of the previous report [11]. The inset in Figure 3 shows the laser ablation energy dependent size variation of CNPs, the color gradually changed from colorless (ethanol liquid medium) into brown in short time intervals, which was confirms the growth of CNPs. In the meantime, the ethanol medium containing CNPs revealed an enhancement in the percentage of absorption intensity with increasing the laser ablation energy (Figure 3). This suggested that the NPs number density, shape and size was indeed influenced by the laser ablation energy. It is value noting that this mechanism of CNPs formation is quite different than other growth methods i.e. vacuum pulse laser deposition technique, where the number and sizes of formed nanoparticles are obvious by the concentration of laser ablated plasma plume on the target surface.

Irrespective of the laser ablation energy, two characteristic bands were evidenced centered around 259 nm and 323 nm (Figure 3). The broadening of these absorption bands verified the size dispersion of nanoparticles grown in the ethanol medium. The intense absorption band was allocated to the phenolic acids and their derivatives (flavanols, cinnamaldehyde, phenylpropenes, and eugenol). The weak absorption peak (Figure 3) revealed a slight broadening accompanied by a tiny redshift. The occurrence of this weak peak was assigned to the presence of the ring of benzoyl and cinnamoyl system in the nucleated CNPs. Furthermore, it is known that the absorption band in the range of 259-263 nm arises due to the present of polyphenols compounds and the protein of plants. This observation indicated the exited of protein into the ethanol solution by cinnamon compound and acted as a probable mechanism for stabilization of CNPs solution. While, the intensity of the prominent absorption bands in the range of 323-327 nm was progressively increased. This observation was ascribed to the quantum size effects of CNPs as the report earlier [11,12].

IOP Conf. Series: Journal of Physics: Conf. Series **1027** (2018) 012002 doi:10.1088/1742-6596/1027/1/012002

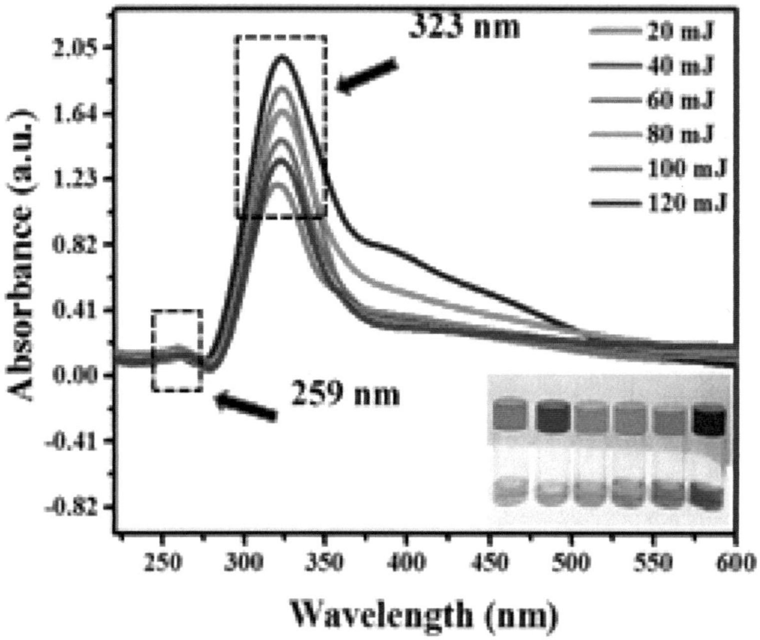

Figure 3. Laser ablation energy dependent optical absorption spectra of CNPs (Inset: particle size dependent color change).

3.3. FTIR Absorption Bands

Figure 4 shows the spectra of all samples in the presence of various functional groups. The structure of compound present in the CNPs and the purity of the liquid medium (ethanol) was analyzed in terms of wavenumbers. The FTIR positions of absorption band spectra of CNPs samples exhibited a slight shift with a change in the absorption band intensity. The appearance of a broad and strong absorption band of CNPs around 3352 cm^{-1} was assigned to the stretching vibration of (O-H) of alcohols and phenols. Whereas, the absorption band located at 2877 cm^{-1} was corresponded to the stretching vibration of (C-H) alkane groups [11]. The absorption band at 2510 cm^{-1} and 1910 cm^{-1} showing the occurrence of -C≡C- the stretch vibration bands of aldehydes and alkynes groups. The gradual decrease in the absorption intensity of the band around 1910 cm^{-1} was due to the ethanol oxidation and subsequent formation of CNPs. The band at 1651 cm^{-1} showed a nonlinear increase in the intensity which was related with the stretching vibration of C=O groups of alkenes [12]. Herein, the growth of CNPs confirmed and supported by the observed broadening and shift in UV-Vis peaks (Figure 3). The dominant peak as showed in Figure 3 corresponded to the presence of major compound in the cinnamon (cinnamaldehyde, flavonoids and aldehydes) [9,11]. The band at 1408 cm^{-1} was allocated to the bending of C-OH vibration alkane. The revealed of absorption bands at 1025 cm^{-1} which was attributes to the presence of C-O stretching of aliphatic amines caused by shifted to the higher wavenumber with the increase in intensity Besides, the absorption band at 857 cm^{-1} was shifted and the intensity was enhanced with increasing ablation energy. The occurrence of this band was assigned to the alkynes and alkyl halides in the located absorption band of C-H bending suggesting their contribution in the CNPs growth. FTIR spectra peaks as shown in Figure 4 established that the CNPs were covered by plant minor metabolites such as flavonoids, glycosides, phenols, terpenoids, tannins, carboxylic acid, aldehyde and ketone [12-14].

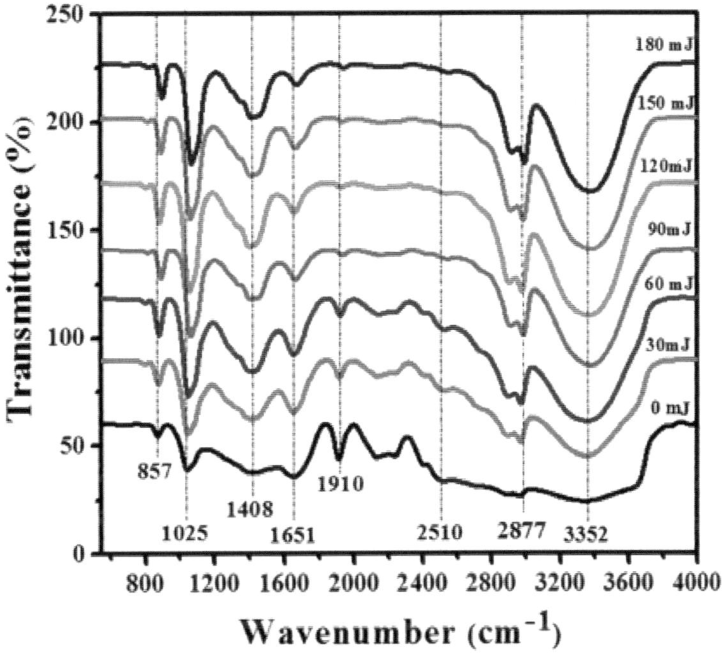

Figure 4. FTIR spectra of CNPs depended on laser ablation energy in the wavenumber range of 500-4000 cm-1.

4. Conclusion

A novel report was to synthesize CNPs immersed in ethanol liquid medium accompany varying laser ablation energy using PLAL technique. The morphology, structure and optical properties of such CNPs were found to be sensitive to the variation of laser ablation energy. UV-Vis spectra revealed two characteristic bands in the range of 259 and 323 nm. The optimum stability of CNPs was attained at laser ablation energy of 80 mJ. Furthermore, the functional groups of CNPs samples are clearly obtained in the range of 900- 1700 cm^{-1} due to the presence of aromatic plants and portions in the nanoparticles. It was established that the proposed natural CNPs may be significantly prospective for the medicine application.

Acknowledgement

Authors are thankful to Malaysian Ministry of Education for the financial assistance through GUP Tier 1 15H47.

References

[1] Smith A M, Duan H, Rhyner M N, Ruan G, and Nie S 2006 A systematic examination of surface coatings on the optical and chemical properties of semiconductor quantum dots. *Physical Chemistry Chemical Physics* **8**(33) 3895-3903

[2] Pearce J M 2012 Physics: make nanotechnology research open-source. *Nature* **491**(7425) 519-521

[3] Makarov V V, Love A J, Sinitsyna O V, Makarova S S, Yaminsky I V, Taliansky M E and Kalinina N O 2014 Green nanotechnologies: synthesis of metal nanoparticles using plants. *Acta Naturae* **6** (20)

[4] Safari J and Zarnegar Z 2014 Advanced drug delivery systems: nanotechnology of health

design a review. *Journal of Saudi Chemical Society* **18**(2) 85-99

[5] Ahmed S, M Ahmad, Swami B L and Ikram S 2016 A review on plants extract mediated synthesis of silver nanoparticles for antimicrobial applications: a green expertise. *Journal of Advanced Research* **7**(1) 17-28

[6] Yang G 2012 Laser ablation in liquids: principles and applications in the preparation of nanomaterials *CRC Press*

[7] Wagener P, Schwenke A, Chichkov B N, and Barcikowski S 2010 Pulsed laser ablation of zinc in tetrahydrofuran: bypassing the cavitation bubble. *The Journal of Physical Chemistry C* **114** 7618-7625

[8] Sathishkumar M, Sneha K and Yun Y S 2010 Immobilization of silver nanoparticles synthesized using Curcuma longa tuber powder and extract on cotton cloth for bactericidal activity. *Bioresource Technology* **101**(20) 7958-7965

[9] Gunawardena D, Karunaweera N, Lee S, van Der Kooy F, Harman D G, Raju R and Münch G 2015 Anti-inflammatory activity of cinnamon (C. zeylanicum and C. cassia) extracts the identification of E-cinnamaldehyde and o-methoxy cinnamaldehyde as the most potent bioactive compounds. *Food and Function* **6** 910-919

[10] Menéndez-Manjón A, Wagener P and Barcikowski S 2011 Transfer-matrix method for efficient ablation by pulsed laser ablation and nanoparticle generation in liquids. *Journal Physics Chemistry C* **115** 5108–5114

[11] Fatima M, Zaidi N U, Amraiz D and Afzal F 2016 In vitro antiviral activity of Cinnamomum cassia and its nanoparticles against H7N3 influenza a virus. *Journal of Microbiology and Biotechnology* **26** 151

[12] Salim A A, Bidin N, Lafi A S and Huyop F Z 2017 Antibacterial activity of PLAL synthesized nanocinnamon. *Materials and Design* **132** 486-495

[13] Salim A A and Bidin N 2017 Pulse Q-switched Nd: YAG laser ablation grown cinnamon nanomorphologies: Influence of different liquid medium. *Journal of Molecular Structure* **11**(49) 694-700

[14] Maruthamuthu R and Ramanathan K 2016 Phytochemical analysis of bark extract of Cinnamomum verum a medicinal herb used for the treatment of coronary heart disease in malayali tribes. *International Journal of Pharmacognosy and Phytochemical Research* **8** 1218-1222

IOP Conf. Series: Journal of Physics: Conf. Series **1027** (2018) 012003 doi:10.1088/1742-6596/1027/1/012003

Double fillet lap of laser welding of thin sheet AZ31B Mg alloy

Mahadzir Ishak* **and M N M Salleh**

Faculty of Mechanical Engineering, Universiti Malaysia Pahang, 26600 Pekan, Pahang

Corresponding author: mahadzir@ump.edu.my

Abstract. In this paper, we describe the experimental laser welding of thin sheet AZ31B using double fillet lap joint method. Laser welding is capable of producing high quality weld seams especially for small weld bead on thin sheet product. In this experiment, both edges for upper and lower sheets were subjected to the laser beam from the pulse wave (PW) mode of fiber laser. Welded sample were tested their joint strength by tensile-shear strength method and the fracture loads were studied. Strength for all welded samples were investigated and the effect of laser parameters on the joint strength and appearances were studied. Pulsed energy (EP) from laser process give higher effect on joint strength compared to the welding speed (WS) and angle of irradiation (AOI). Highest joint strength was possessed by sample with high EP with the same value of WS and AOI. The strength was low due to the crack defect at the centre of weld region.

1. Introduction

Joining process has become a great demand where it is important to make two or more similar or different parts to become one product with good characteristics. There are several types of joining method such as bolt and fasteners, adhesive, solder, rivet, and welding method. In all types of joining, welding gives better results especially in metal joining such as ferrous or non-ferrous metals. It is one of the methods that have been applied in manufacturing industries to joint products crucially towards metal. Manufacturing industries play an important role in term of producing new products and technologies to the people around the world. Transportation, architecture, city development, and others need manufacturing industry in order to accomplish their desired products. The application of thin sheet metals usually found in automotive, aviation, and electronic devices as an alternative of reducing weight [1, 2]. Thin sheet metals has been a growing demand in manufacturing engineering where the thickness used is below than 2 mm [3-5]. Among the metals used in industries, magnesium alloys such as AZ31B, AZ91D and AZ61 has increasing application due to the advantages; light weight metal, high-strength to weight ratio, good weldability, high electrical conductivity and high corrosion resistance [6, 7].

Application of laser in material processing is increasing since it was introduced from 1960 [8]. This contribution includes the laser welding process to join metals where this method can be used even to weld thin sheets metals. The metals to be welded are melted by means of heat from the laser beam which focused at one spot where it creates weld pool. Laser welding is unique compared to the conventional welding that have been applied in industries long before. It is unique because of its advantages as a bonding process. The advantages includes producing powerful laser beam, concentrated heat source, and allowing to obtain deep and narrow penetration depth for a high speed

IOP Conf. Series: Journal of Physics: Conf. Series **1027** (2018) 012003 doi:10.1088/1742-6596/1027/1/012003

process [9]. This process already becomes an important manufacturing process especially in joining metals due to its advantages as a bonding process.

As for industrial requirements, fiber laser welding is promising bigger advantages for manufacturing industry such as low cost, time saving, free maintenance and high quality production [10]. In this paper, thin sheet AZ31 was welded by fiber laser welding method with double fillet lap joint type was preferred. The objective of this work is to investigate the parameters used to achieve better weld joint in term of its strength and weld appearances.

2. Methodology/Model Description

2.1 Material

AZ31B Magnesium alloy with thickness of 0.6 mm was used in this study. The major alloying element for this material is aluminium (Al), and zinc (Zn). Metal composition was Al-2.8 %, Zn-1.2 % and Mg- balance. Before the metal subjected to the laser welding, it was cut by dimension 53 × 20 × 0.6 mm. Table 1 shows the chemical compositions of AZ31B which tested with spectrometer and from previous research. Samples were polished with 180 grit sand paper in order to remove the oxide layer presented at the welding area.

Table 1. Chemical composition of AZ31B

Element	Weight, %		
Al	2.8	2.98	2.46
Zn	1.2	0.87	1.7
Mn	0.4	0.35	0.58
Si	0.02	0.016	<0.1
Cu	0.002	0.004	<0.05
Fe	0.002	0.001	<0.005
Mg	Bal.	Bal.	Bal.
Source	Spectrometer	F. Pan, et al., 2017 [2]	M. Harooni, et al., 2015 [11]

2.2 Welding Setup

Pulse wave (PW) mode was used in this experiment where the pulsed energy, EP used are 1.8, 2.0, and 2.2 J with constant pulse width and pulse repetition rate of 2 ms and 60 HZ, respectively. The angle of irradiations used is 2, 3, and 4 degrees and the welding speed used is 2, 3, and 4 mm/s. Fifteen samples were welded in this experimental work and the bond width, throat length, and penetration depth of all samples were measured and collected. Figure 1 (a) shows a low power fiber laser machine with the average power of 200 W used in this research work.

Welding jig was fabricated in order to perform the lap joint welding where the jig was made by Aluminium. Welded sample would be clamped on the welding jig where the spacer was compulsory for the lap joint sample. Arduino system was adapted to the welding rail where it controlled the movement of the stepper motor for the x-axis (welding line). Shielding gas was then attached to the laser head where the nozzle could be adjusted to the suitable angle for different welded metal as shown in Figure 1 (a) and (b) for the welding setup for the double fillet welding technique. Shielding gas used was Nitrogen (N_2) with flow rate 25 L/min.

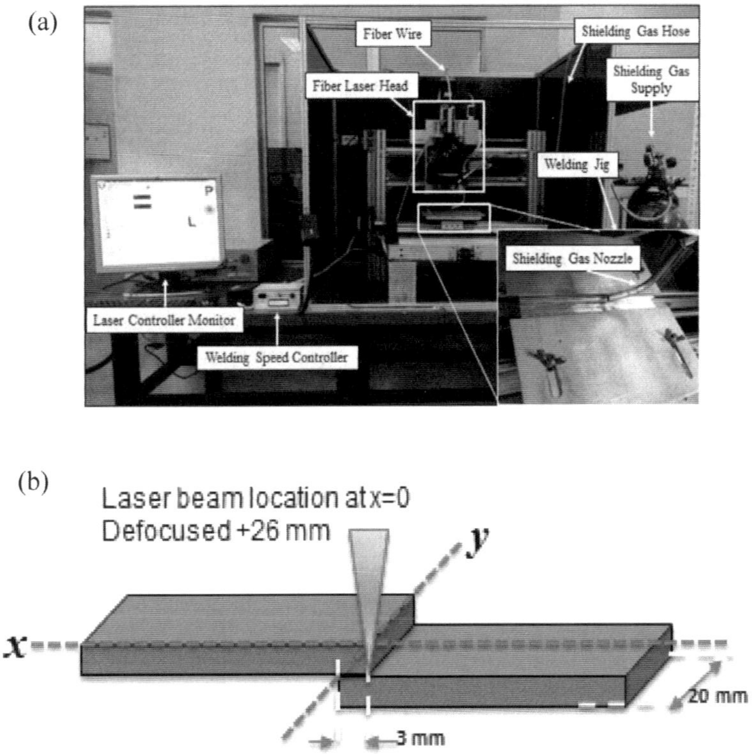

Figure 1. Double fillet lap laser welding condition (a) Complete welding setup and (b) Schematic illustration during welding

3. Result and Discussion

3.1 Verification

Laser welding experiment was carried out on thin sheet of AZ31B Mg alloys according to the number of run order as shown in Table 2. Average values of tensile-shear strength for each tested sample were tabulated where it showed the average result of fracture load values which have been conducted for all fifteen welded samples. It was observed that specimen number 1 has the lowest tensile-shear strength of 22.3 MPa with a fracture load of 101.5 N produced by Ep: 1.8 J, WS: 2 mm/s, and AOI: 3°. The joint produced by Ep: 2.0 J, WS: 2 mm/s and AOI: 2° (sample 9) achieved the highest tensile-shear strength (62.0 MPa). Sample 8 showed the second highest of shear strength (52.0 MPa) among others which was 10 MPa difference to the strength of sample 9. Sample 13 to 15 showed approximately same results of strength which was within 30 to 32 MPa. However, sample 13 possessed a slightly higher fracture load with 520 N compared to sample 14 and 15 with 426.3 N and 450 N, respectively. Figure 2 presents the bar graph in order to show the average values of fracture loads possessed by each welded sample. The error bar represents the collected values from high and low value of each sample as shown in Table 2. Based on Figure 2, it was observed that sample 9 showed the highest fracture load of 740.2 N with 62 MPa shear strength, meanwhile sample 1 showed the lowest fracture load of 101.5 N with 22.3 MPa. Based on the graph, it was observed that the welded samples possessed the fracture load mostly ranged between 400 to 600 N.

Table 2. Fracture load result according to experiment order

Sample	EP (J)	WS (mm/s)	AOI (°)	Fracture Load (N)		
				Low	High	Avg
1	1.8	2	3	91.50	111.5	101.5
2	2.2	2	3	574.8	604.8	589.8
3	1.8	4	3	180.5	210.5	195.5
4	2.2	4	3	467.7	497.7	482.7
5	1.8	3	2	242.6	272.6	257.6
6	2.2	3	2	619.3	639.3	629.3
7	1.8	3	4	505.5	525.5	515.5
8	2.2	3	4	534.1	564.1	549.1
9	2.0	2	2	730.2	750.2	740.2
10	2.0	4	2	578.4	608.4	593.4
11	2.0	2	4	277.1	307.1	292.1
12	2.0	4	4	165.4	195.4	180.4
13	2.0	3	3	510.0	530.0	520.0
14	2.0	3	3	416.3	436.3	426.3
15	2.0	3	3	435.0	465.0	450.0

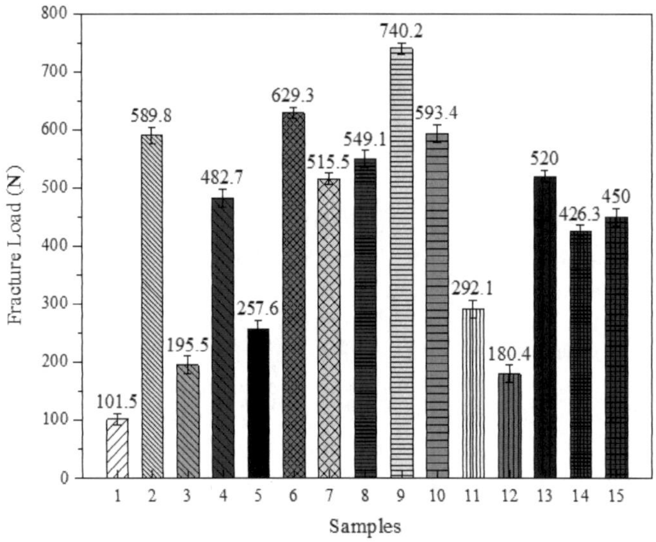

Figure 2. Average fracture load values for 15 welded samples

Sample 9 possessed good weld with less occurrence of defect compared to other samples. It was proved with the joints' strength in Table 2 where sample number 9 had the highest strength among others. Cracks occurred in most of the weld due to the hot crack which was generated when low welding speed was used. Much higher energy input caused distortion effect on the sheets which then produced higher stress concentration between the two sheets [12].

The crack in the weld bead occurred also was inevitable due to the two sheets of the welded was not clamped properly before the welding process which could present the gap between the sheets. Most of the crack defects were found at the second weld area except sample 2, 6, 9, 11, 13, 14, and 15.

Sample 13 to 15 shows approximately same weld bead for both fillets as these samples were welded using same laser parameters.

Sample number 1 was found as the lowest joint strength as the crack defect presented at the second weld region referred from Figure 4.8 (a). There was a high possibility that the crack was produced due to the solidification crack occurred after being welded using laser which created rapid cooling process in the fusion zone [12]. In addition, AZ31B consist of Zinc, Zn element which have wide melting range where it is sensitive to hot cracking [13].

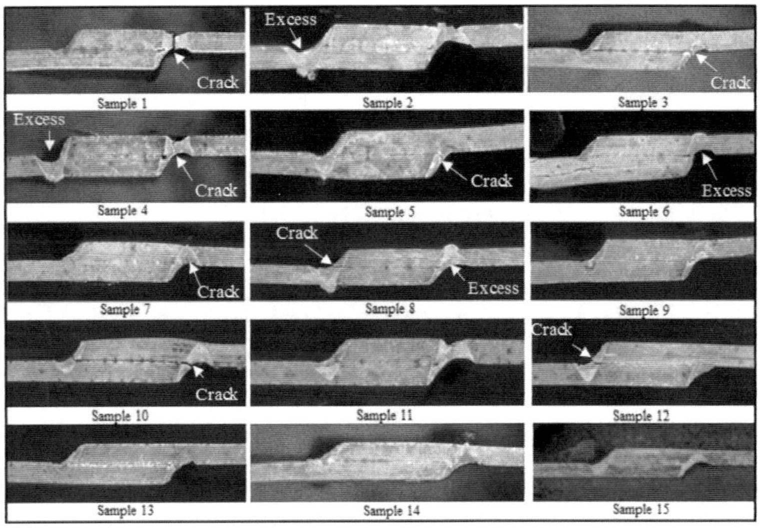

Figure 3. Macrostructure of 15 welded samples

4. Conclusion

From the present study, the followings can be concluded:

(a) Higher pulsed energy with same value of WS and AOI produced higher joint strength.
(b) Highest pulsed energy produced sound weld and free defects based on its appearance compared to the lower pulsed energy
(c) Pulsed energy (EP) was the most effective parameter towards joint strength

Acknowledgement

The authors express gratitude to the Malaysian Ministry of Education (MOE) and Universiti Malaysia Pahang in supporting this experimental work with grant GRS150345.

References

[1] Simoncini M and Forcellese A 2012 Effect of the welding parameters and tool configuration on micro- and macro-mechanical properties of similar and dissimilar FSWed joints in AA5754 and AZ31 thin sheets *Materials & Design* **41** 50-60
[2] Pan F, Zeng B and Jiang B 2017 Enhanced mechanical properties of AZ31B magnesium alloy thin sheets processed by on-line heating rolling *Journal of Alloys and Compounds* **693** 414-420
[3] Steglich D, Tian X, Bohlen J, *et al.* 2014 Mechanical Testing of Thin Sheet Magnesium Alloys in Biaxial Tension and Uniaxial Compression *Experimental Mechanics* **54** 1247-1258
[4] Falconnet E, Chambert J, Makich H, *et al.* 2015 Prediction of abrasive punch wear in copper alloy thin sheet blanking *Wear* **338-339** 144-154

[5] Wu H Y, Sun P H, Zhu F J, *et al.* 2012 Tensile Properties and Shallow Pan Rapid Gas Blow Forming of Commercial Fine-grained Mg Alloy AZ31B Thin Sheet *Procedia Engineering* **36** 329-334
[6] Tan Q, Atrens A, Mo N, *et al.* 2016 Oxidation of magnesium alloys at elevated temperatures in air: A review *Corrosion Science* **112** 734-759
[7] Molnár P, Ostapovets A and Jäger A 2014 Reversible motion of twin boundaries in AZ31 alloy and new design of magnesium alloys as smart materials *Materials & Design* **56** 509-516
[8] Odabaşı A, Ünlü N, Göller G, *et al.* 2010 A Study on Laser Beam Welding (LBW) Technique: Effect of Heat Input on the Microstructural Evolution of Superalloy Inconel 718 *Metallurgical and Materials Transactions A* **41** 2357-2365
[9] Moskvitin G V, Polyakov A N and Birger E M 2013 Application of laser welding methods in industrial production *Welding International* **27** 572-580
[10] Assunção E, Quintino L and Miranda R 2009 Comparative study of laser welding in tailor blanks for the automotive industry *The International Journal of Advanced Manufacturing Technology* **49** 123-131
[11] Harooni M, Ma J, Carlson B, *et al.* 2015 Two-pass laser welding of AZ31B magnesium alloy *Journal of Materials Processing Technology* **216** 114-122
[12] Ishak M, Yamasaki K and Maekawa K 2012 Laser Welding of Thin Sheet Magnesium Alloys In *Nd:YAG Lasers* InTech Open Access **2016** 135-157
[13] Yu Z H, Yan H G, Yin X Y, *et al.* 2012 Liquation cracking in laser beam welded joint of ZK60 magnesium alloy *Transactions of Nonferrous Metals Society of China* **22** 2891-2897

IOP Conf. Series: Journal of Physics: Conf. Series **1027** (2018) 012004 doi:10.1088/1742-6596/1027/1/012004

Fiber-based optic sensor for detecting human blood clot: present and future revival

Nada Elshikeri and Hazri Bakhtiar*

Faculty of Science, Universiti Teknologi Malaysia, 81310, Skudai, Johor Bahru, Johor, Malaysia.

Corresponding author: hazri@utm.my

Abstract. Sustaining human's life-frame away from being impeded by the clot - ghost term, we attempt to approach a mobile fiber-based optical sensor (f-s) for detecting blood clot in a blood vessel (intra-arteries/veins). Blood vessels are the part of the circulatory system that transport blood throughout the human body, thus their significance of being protected arise to the monograph focus. MRI (magnetic resonance imaging), X-rays and other medical instruments are diagnostic immobility techniques with a slackest interval. The corer causation of fiber-based optical sensor is to detect a clump of blood in the bloodstream by providing a prompt mobile diagnostic intervals preserving last-minutes-breath of human's life. The detector (f-s) has been etched by diluting sulphuric acid ~10% at certain zone to sensate its function. The in-vitro monograph peaks its maximal monitoring when the sensor is attached to Raman Spectroscopy (RS) setup. RS quantifies the relative intensities of fibrinogen bond, which is the first type of blood coagulation elements of blood plasma. Blood coagulation parameters are the major concern of the monograph investigation, such as total haemoglobin (tHb), clotting reaction time (t), clot progression time (t2), maximum clot amplitude (ma) and mean refractive index (r). A blood sample will be drawn from the patient and after centrifugation to separate blood plasma from its constituents, then an immediate sloshing of blood plasma in the (f-s) packet which has its plug-in to RS. Estimating the quantitative analysis of blood sample concentration, RS will determine the presence of coagulation in terms of intensity and medical procedures will dominate the treatment process. Thus, the suggestive monograph provides a definite instrument for investigating blood coagulation intra-arteries/veins promptly.

1. Introduction

Blood is the vital requirements of our body which supplies oxygen, nutrients, remove waste products and versatile positive functions. It includes plasma (the liquid portion), blood cells (which come in both red and white varieties), and cell fragments called platelets. Cells and platelets make up about 45% of human blood, while plasma makes out the other 55%. Blood plasma serves a variety of functions, from maintaining a satisfactory blood pressure and volume to supplying critical proteins from blood clotting and immunity. It also serves as the medium for exchange of vital minerals such as sodium and potassium and helps to maintain proper(acid-base) balance in the body, which is critical to cell function. Plasma is obtained by separating the liquid portion of blood from the cells. The combination between blood and clot has its quite specific scenario. One scenario, apparently, when the lining of a blood vessel is damaged, platelets are attracted to the wound site, where they form a sticky plug.

The platelets release signals, which not only attract other platelets and make them become form soluble protein present, a water fibrinogen cascade that ultimately converts signalling sticky, but also activate a signalling cascade that ultimately convert fibrinogen, a water-soluble protein present blood plasma, into fibrin (a non-water soluble protein). The fibrin forms threads that reinforce the platelet plug, making a clog that prevent further loss of blood [1,2]. According to biochemistry, the different molecules of the fibrinogen are connected by the peptide bombs and then fibrinogen clots come in into being [11]. The hazardous is the bloodstream clot which may have potentially significant consequences. Coagulopathy, a condition in which blood coagulation is impaired, can result from a variety of conditions including: obesity, pregnancy, immobility (including prolonged inactivity, long trips by plane or car), smoking, oral contraceptives, certain cancers, trauma, age (increased risk for people over age 60), a family history of blood clots, chronic inflammatory diseases, lack of physical activity, diabetes, high blood pressure and high cholesterol.

Intrinsically, optical fiber-based sensor (f-s) imprint its growing importance as an advance biomedical instrumentation to enable more efficient patient diagnosis, monitoring, and treatment. F-s is extremely immune to electromagnetic interference (EMI), chemically inert, nontoxic, and electromagnetic and radio frequency (RF) signals. Thus, it's use will not cause interference with the conventional electronics found in medical theatres as well as during thermal ablative treatments involving RF or microwave radiation [3].

Biomedical (f-s) can be categorized into four main types: physical, imaging, chemical, and biological. Biological sensor is the major focus in the monograph whereas it tends to be more complex and rely on biologic recognition reactions-such as enzyme-substrate, antigen-antibody, or ligand-receptor which to identify and quantify specific biochemicals molecules of interest [4,5]. As a rise of (f-s) dramatically in the biomedical domain, thus it becomes a giant option to detect blood clot (coagulation) precisely. The present invention relies on the finding that blood clot in a blood vessel generate a unit and specific spectrum detectable by Raman spectroscopy [6].

In particular, Raman spectroscopy in the near infrared (NIR) range measures vibrational transitions in molecules, so it can detect structural and clinical medical character of material in molecules in molecular level. In addition, its signal is more intense in high wave number region than in fingerprint region which consequently contribute to further improving the sensitivity to tissue pathological changes [7,8].

2. Experimental Setup and Data Processing

Fiber-based optical sensor applied for the monograph has been etched by diluting sulphuric acid (etching) ~ 10%. The packet for the blood collection demonstrates the sensing zone that had been etched by the etchant as in Figure 1.

Raman scattering happens wherein a laser photon of well-defined energy is scattered off a molecule, a small amount of energy is lost to a molecular vibration so the spectrum of the scattered light shows its sharp peak spectral features, characteristics for each molecule. In our monograph, we will obtain Raman spectrum of normal and suspected person's blood plasma using FT (Fourier Transform) -Raman spectrometer with the exciting wavelength 1064nm.

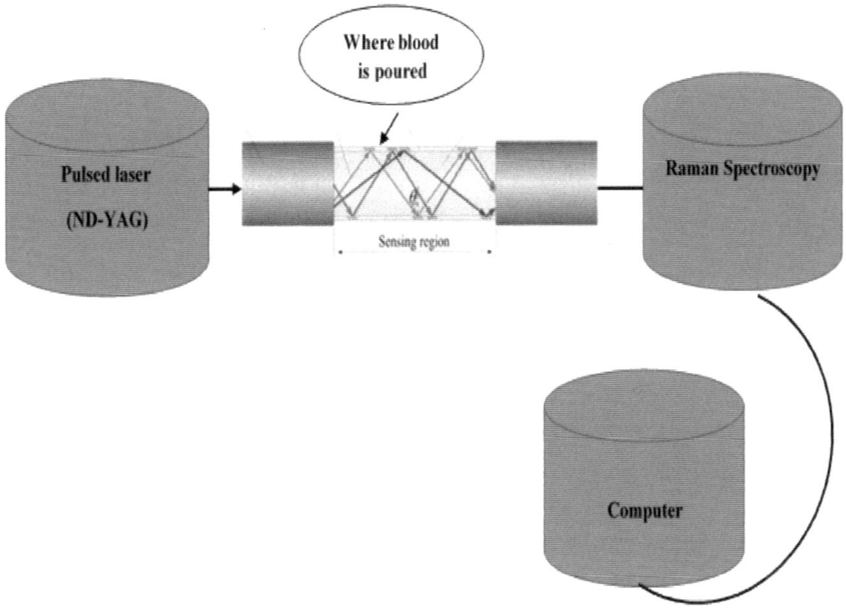

Figure 1. Experimental setup

The major distinction of blood plasma is its relative intensities, especially in the wavenumbers of 800-950cm^{-1} and 1530-1570cm^{-1}. The near-infrared (1064nm) excitation offers higher powered lasers to be used without photo degradation of the sample. Blood plasma is prepared by spinning a tube of fresh blood containing an anticoagulantin a centrifuge until the blood cells fall to the bottom of the tube. The blood plasma is then poured or drawn off [12]. Blood plasma has normal density of approximately 2-4g/L. If the fibrinogen in the blood reaches a certain level (That is to say, if fibrinogen is more than 2-4g/L), then the peptide bonds of fibrinogen would possibly change into some stable fibrinogen clots [13].

With this exertion, conformational and structural features of human blood will be studied by spectroscopic techniques with particular reference to relative intensities of fibrinogen peptide bond [9,10]. As well as featuring the monograph parameters such as total haemoglobin (tHb), clotting reaction time (t), clot progression time (t2), maximum clot amplitude (ma) and mean refractive index (r). Samples of whole blood will be drawn from different gender of average age about 50.0 ± 12.6.

3. Results

Figure 2 (a) shows the reference concentrations of fibrinogen, coagulation time, clot progressing time and time expected wavenumber for coagulation, (b) and (c) predict that the clotting might take place in Raman shift below 2500cm^{-1} with the sharpest peak. Fiber-based optical sensor facilitates the sensation zone in the range less than the reference concentration through the predicted sharpest peaks.

IOP Conf. Series: Journal of Physics: Conf. Series **1027** (2018) 012004 doi:10.1088/1742-6596/1027/1/012004

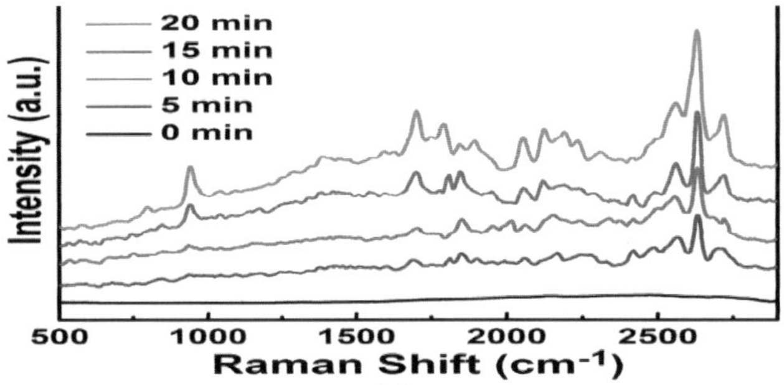

(a)

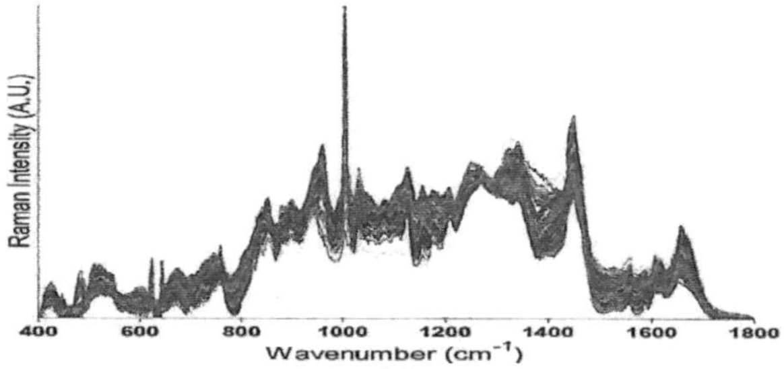

(b)

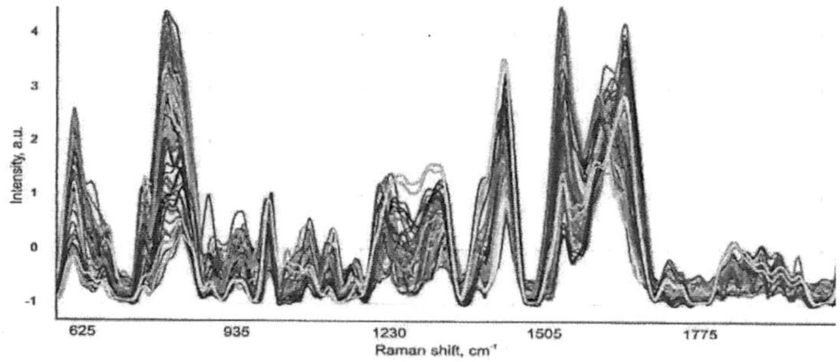

(c)

Figure 2. (a) Raman shift below predict that the clotting might take place and (b) and (c) are the wavenumber for coagulation

4. Conclusion

The method provides an opportunity to detect abnormal fibrinogen in terms of intensities. Hence it appears that the fiber-based optical sensor and Raman spectroscopic method will show a high potential for analysis of blood plasma in the whole blood. The drawback of those systems (F-s and Raman spectroscopy) is that although they may identify a vessel obstruction (e.g. blood clot), they are unable to localize the obstruction. As an expectation, F-s is a highly reproducible technique in the detection of clot signals in a complex biomedical matrix. The method and instrumentation demonstrated in the monograph are under great prediction and observation of present and future revival.

5. Ethical and Consideration

An official letter will be issued from University Technology Malaysia to demonstrate the blood drawing in clinics or hospitals wherein Raman spectroscopy is employed. Furthermore, revealing genetic information has important ethical implications for individuals as family members. The way that a society governs the disclosure of such information and the extent to which its laws or other regulatory frameworks control what can be disclosed. For instance, we ought, ethically, to seek consent from people to use their genetic information in research because doing so respects their autonomy and freedom to choose.

References

[1] Components of the blood, by OpenStax College, Biology (CC BY 4.0)
[2] Production of the formed elements, by OpenStax College, Anatomy & Physiology (CC BY 3.0)
[3] E Udd 1988 Overview of Fiber Optic Applications to Smart Structures *Review of Progress in Quantitative Nondestructive Evaluation* Plenum Press
[4] B Culshaw and J Dakin Eds. 1989 *Optical Fiber Sensors: Systems and Applications, Vol. II*, Artech House
[5] J I Peterson and G G Vurek 1984 Fiber-Optic sensors for biomedical applications *Science* **224** (4645) 123-127
[6] Continuation-in-part of PCT/US2006/061742 filed in Dec. 7 2006 *Provisional application* **60** (748) 289
[7] N Uzunbajakava, A Lenferink, Y Kraan, E Volokhina, G Vrensen, J Greve and C Otto 2003 Nonresonant confocal Raman imaging of DNA and protein distribution in apoptotic cells *Biophys, J.* **84** 3968-3981
[8] S Koljenovic, T C B Schut, R Wolthuis, B d Jong, L Santos, P J Caspers, J M Kros, G J Pupples 2005 Tissue characterization using high wave number Raman spectroscopy *J. Biomed. Opt.* **10** 1-11
[9] Cataliotti R and Gresele P 1983 Differences between human platelets and erythrocyte membranes A spectroscopic study with particular reference to IR and laser Raman spectroscopy *Photobiochem. Photobiophys.* **6**(3)159-169
[10] Gao Ti-yu and Ci Yun-xian 2001 Fourier Transform Infrared Spectroscopic Investigation of Human Blood cells in Bone Marrow from AML and CMI Patients *Acta Scientiarum Naturalium Universitatis Pekinensis* **3** 192-196
[11] Lubert Stryer 1980 Biochemistry *W.H.Freeman and Company* 133-135
[12] Maton, Anthea, Jean Hopkins, Charles William McLaughlin, Susan Johnson, Maryanna Quon Warner, David LaHart and Jill D Wright 1993 Human Biology and Health *Englewood Cliffs, New Jersey, USA: Prentice Hall.* ISBN 0-13-981176-1
[13] Wei youn 1995 Bioorganic Chemistry, publications of world book company 329-330

Plasma diagnostic by optical emission spectroscopy on reactive magnetron sputtering plasma –A Brief Introduction

Soo Ren How[1], Nafarizal Nayan[*1], Jais Lias[1], Mohd Khairul Ahmad[1], Mohd Zainizan Sahdan[1,2], Mohamad Hafiz Mamat[3], Mohamad Rusop Mahmood[3], Ali Aldalbahi[4]

[1]Microelectronic and Nanotechnology- Shamsuddin Research Centre (MiNT-SRC) & Faculty of Electrical and Electronic Engineering, Universiti Tun Hussein Onn Malaysia (UTHM), Batu Pahat,86400 Johor, Malaysia
[2]Preston GeoCEM Sdn Bhd (1202204-H), No 33 A, Jalan Universiti 4, Taman Universiti, 86400 Parit Raja, Johor, Malaysia
[3]NANO-ElecTronic Centre (NET), Faculty of Electrical Engineering, Universiti Teknologi MARA (UiTM), 40450 Shah Alam, Selangor, Malaysia
[4]Chemsitry Department, King Saud University, Riyadh 11451 Saudi Arabia

Corresponding author: nafa@uthm.edu.my

Abstract. Optical emission spectroscopy has been employed as non-invasive plasma diagnostics method and it is to detect the light emission during titanium nitride deposition using magnetron sputtering plasma discharge. The optical emission from argon and titanium were used to observe the ionization ratio of each element during the discharge. Then, the optical emission of nitrogen spectrum was used for the calculation of gas rotational temperature by using a simple formula and it was evaluated at various discharge conditions. A comparative of two different spectrometers resolution have been used for the determination of gas temperature. The results showed that the evaluated temperatures were almost the same as expected in high-pressure magnetron sputtering plasmas even when using low-resolution of spectrometer.

1. Introduction

Reactive magnetron sputtering is a common technique in microelectronic industries to fabricate oxide and nitride thin films. This technique produced thin films with good quality of adhesion and surface morphology [1-3]. The other advantage of reactive magnetron sputtering technique is that one could control the composition of deposited thin film by controlling the composition of sputter target and reactive gases introduce into the discharge chamber. During the fabrication of integrated circuit, titanium nitride thin film is used as barrier layer to prevent leakage current of the interconnects.

In order to optimize the plasma processing, it is essential to diagnose the plasma behaviour and understands the fundamental process during the thin film deposition. There are several of plasma diagnostics techniques have been discovered for many years. Optical emission spectroscopy is one of the plasma diagnostic tools that has been used for industrial applications [4-8]. In addition, optical emission spectroscopy (OES) measurement has been recognized as one of the common plasma diagnostics techniques due to its simplicity and low cost. Subsequently, another benefit of using OES diagnostics is that one could able to acquire the data without perturbing the plasma during the diagnostic process.

2. Experimental Setup

Reactive magnetron sputtering plasmas were produced using water-cooled conventional balance sputter source. The sputter target was titanium and the discharge gas was argon. Nitrogen was introduced into the chamber as reactive gas to produce titanium nitride thin film. The size of solid titanium sputter target is 75 mm in diameter and 15 mm in thickness. The size of the chamber is 1 m in diameter and 1 m in height. Sputtering chamber consist of six view ports using quartz glass window for plasma monitoring. The vacuum was created using turbo molecular pump backed by rotary pump. Maximum vacuum condition was achieved at 8 x 10^{-6} Torr.

Figure 1 shows the experimental setup to capture the optical emission from sputtering plasmas during the deposition of titanium nitride thin film. An optical lens was used and placed in front of the quartz window chamber as shown in Fig. 1. The purpose of this configuration is to ensure that the emission from specific position in plasma discharge is focused onto the fiber entrance. The ultraviolet and visible emission from the plasma discharge passed through a quartz window chamber. The spectrometer from Ocean Optics HR4000 was used to capture the emission from the plasma discharge. The system is microcontroller-controlled spectrometer and all the operating parameters are implemented through Spectra Suite software interfacing to the unit. The spectrometer able to responsive to the wavelength in the range of 200 nm to 1100 nm. The optical emission of titanium, argon and nitrogen were studied at various of discharge conditions.

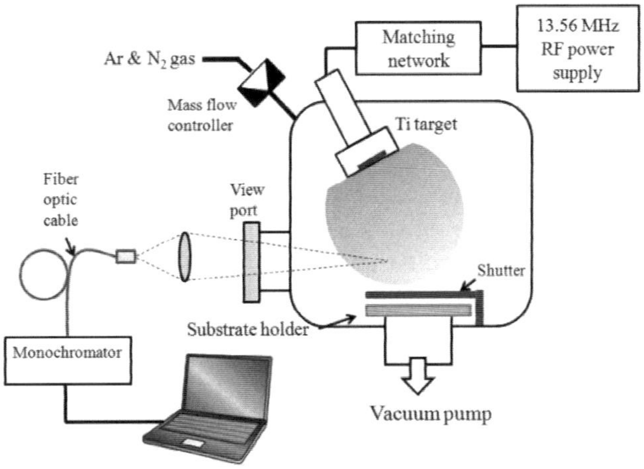

Figure 1. Experimental setup for optical emission spectroscopy measurement at specific position in magnetron sputtering plasma system.

3. Results and Discussion

The discharge power and argon flow rate was kept at 200 W and 36 sccm, respectively. The total gas pressure was kept high at 100 mTorr. The major reason of this high-pressure condition is to enhance the ionization ratio of titanium atom in the sputtering plasma as we have presented it elsewhere [9].

Figure 2 shows the typical emission for magnetrpn sputtering plasma employing titanium target [10]. The nitrogen flow was varied between 0 and 40 sccm. The assingment of the elemental emission spectrum was confirmed from NIST database [11]. It is clearly shown that the emission from titanium atom and ions between 300-500 nm are significant when no the nitrogen flow into the chamber (nitrogen flow rate = 0 sccm). The emission from argon metastables and ions between 700-850 nm are also clearly shown in Figure 2.

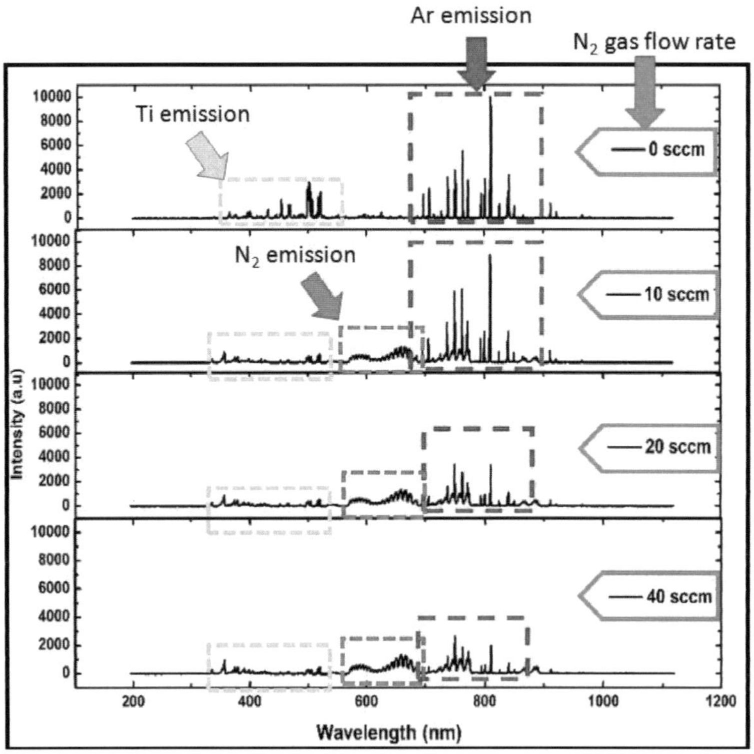

Figure 2. Widerange emission of magnetron sputtering plasma using titanium target and various nitrogen flow rates[9].

Figure 3 shows the emission of argon atoms and ions that has been assigned as Ar I and Ar II, respectively. As shown in Figure 3, the ratio of Ar I / Ar II decreased with the nitrogen gas flow rate. This result indicates two situations: (1) the relative intensity of Ar II increased with the nitrogen gas flow rate and (2) the relative intensity of Ar I decreased with the nitrogen gas flow. The experimental result was not consistent with the global model simulation on titanium nitride deposition using reactive magnetron sputtering plasmas [12-14]. The main reason of this inconsistency is may be due to the high-pressure condition in the present experiment. To date, there is no simulation report on the high-pressure magnetron sputtering plasmas.

Figure 4 show the relative intensities of titanium atom (Ti I) and titanium ion (Ti II) at various nitrogen flow rates. As shown on Figure 4, the characteristics of titanium emission was almost the same with argon emission. The Ti II increased with the nitrogen flow rate. This understanding is very important for titanium nitride thin film deposition using ionized physical vapor deposition technique. In general cases, the electron-impact excitation from titanium ground states of species is considered to be dominant excitation mechanism to increase the density of titanium ions in the plasmas. The intensity of Ti I is directly proportional to the concentration of titanium atoms sputtered from the cathode.

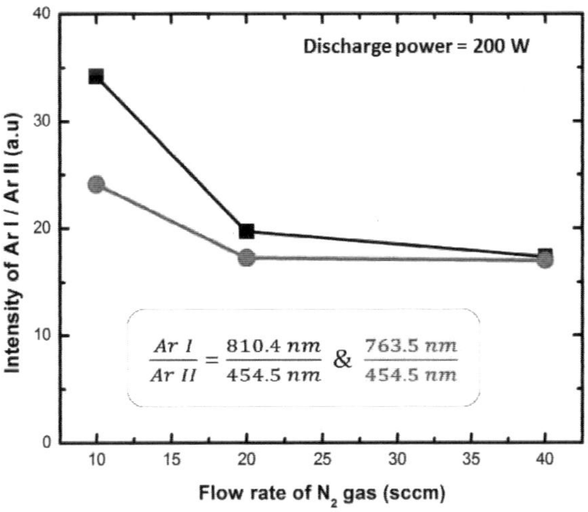

Figure 3. Emission ratio of argon atom and argon ion at various nitrogen flow rates.

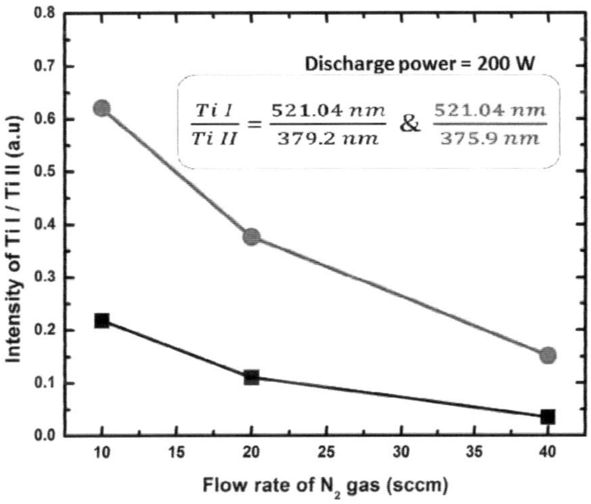

Figure 4. Emission ratio of titanium atom and titanium ion at various nitrogen flow rates.

Another advantage of optical emission spectroscopy is to evaluate plasma gas temperature from the rotational temperature of nitrogen gas. The determination of rotational temperature of nitrogen gas was done by using any rotational band of the emission spectrum of nitrogen and it is applicable for any other molecular gas such as oxygen and hydrogen. In the case of nitrogen, one could use either the band from the first negative system at 391.4 nm or second positive system bands at 337.4 or 380.4 nm. The spectrum calculation procedure for molecular gas temperature is based on the basic principles of optical emission and it requires knowledge of the electronic, vibrational and rotational constants of particular molecular spectrum. In our previous work, data of the nitrogen spectrum was captured at 775 nm as shown in Fig.5 [15]. For simplicity, N.Britun *et al* in his previous works had calculated the rotational temperature based on the theoretical formula after some correlation of experimental data [16]. The formula is given by:

$$Trot\,[K] = \frac{195}{\left({}^{P1}/_{P2}\right) - 0.52} \quad (1)$$

Figure 5 show emission of nitrogen spectrum measured using two different type of spectrophotometers. The objective was to compare the evaluation of gas temperature when measured using low- and high-resolution spectrophotometers. If the results are comparable, one should able to use simple and low cost spectrophotometer for gas temperature evaluation. As shown in Figure 5, the peak at two positions that required for equation 1 was clearly detected for both low- and high-resolution spectrophotometers.

Figure 6 shows the nitrogen gas rotational temperature evaluated from Figure 5 and equation 1. The rotational temperature was evaluated at plasma discharge power of 100W, 150W and 200W. As shown in Fig. 6, the temperature increased with the discharge power. The fluctuation of gas temperature evaluated from low- and high- resolution spectrophotometers were small, indicating that the gas temperature at the range of 410-460 K. In other words, by using low resolution spectrometer is sufficient to detect the nitrogen emission spectrum and peak values that are required to calculate the nitrogen gas rotational temperature. In addition, Figure 6 indicates that the depositon enviroment of titanium nitride thin film was at low temperature. Therefore, the process parameters of the present experiment are suitable for coating of heat sensitive materials such as polymers.

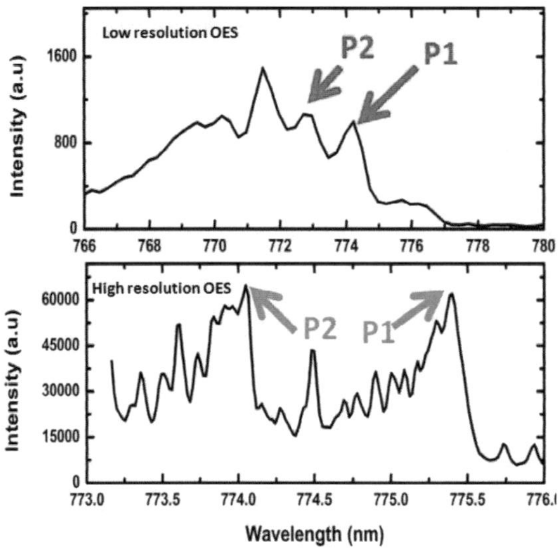

Figure 5. Nitrogen emission captures using low- and high-resolution spectrophotometer[15].

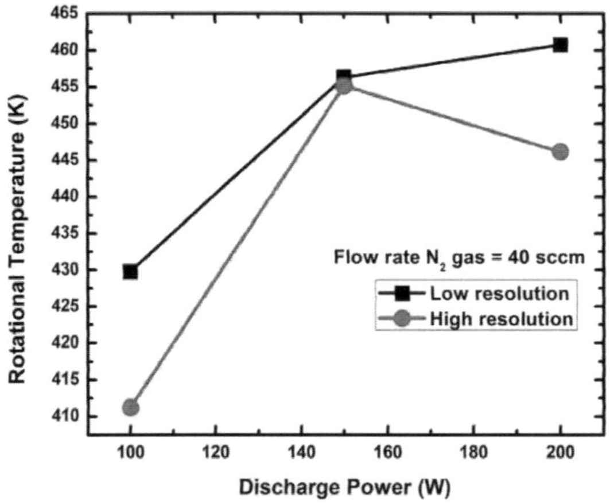

Figure 6. Rotational temperature of nitrogen gas evaluated from the emission spectrum of nitrogen[15].

4. Conclusion

In conclusion, plasma diagnostic using optical emission spectroscopy technique is a simple method to understand the fundamental of plasma processing applications. The ratio of ions and atoms emission peaks were used to evaluated the ionization process at various discharge conditions. The changes were significantly observed. Then the optical emission have been detected and performed by using different resolution of spectrometer. Variation of plasma discharge powers at 100 W, 150 W and 200 W using low-resolution and high-resolution spectrometers were reported. The nitrogen gas temperature evaluated using both spectrometers were almost the same at the range of 410-460 K. The influence of discharge power is almost negligible. Therefore, we concluded that low-resolution spectrometer is adequate when measuring gas rotational temperature in certain region.

Acknowledgement

The present work was partially supported by Contract Research Grant Scheme UTHM-U552 and Ali Aldalbahi acknowledges the financial support by Deanship of Scientific Research, College of Science Research Center, King Saud University.

References

[1] R Pintaske, T Welzel, M Schaller, N Kahl, J Hahn and F Richter 1998 Spectroscopic studies of a magnetron sputtering discharge for boron nitride deposition *Surf. and Coatings Tech.* **99** 266-273

[2] S P Koirala, I Awaah, S L Burkett and M H Gordon 2011 Correlation of plasma characteristics to etch rate and via sidewall angle in a deep reactive ion etch system using Langmuir probe and optical emission spectroscopy *Jour. of Vac. Sci. and Technol.* **29** 1-6

[3] A J M Mackus, S B S Heil, E Langereis, H C M Knoops, M C M van de Sanden and W M M Kessels 2009 Optical emission spectroscopy as tool for studying, optimizing and monitoring plasma-assited atomic layer deposition processes *Jour. of Vac. Sci. & Technology* **28** 77-87

[4] K E Evdokimov, Maxim E Konischev, Vladimir F Pichugin, Z Sun 2017 Study of argon ions density and electron temperature and density in magnetron plasma by optical emission spectroscopy and collisional-radiative model *Resource-Efficient Technologies* **3** 187-193

[5] M K Sharma and B K Saikia 2008 Discharge conditions and emission spectroscopy of N2 and N2+ active species in a variable power dc pulsed plasma used for steel nitriding *Ind. Jour. of Pure and App. Physics* **46** 463-370

[6] F Perry, B Stauder, G Henrion, Ph Pigeat 1995 Plasma diagnostics for the control of reactive magnetron deposition process Surf. and Coatings Tech. **74-75** 575-579

[7] Irving P Herman 2003 Optical diagnostics for thin film processing *Annu. Rev. Phys. Chem.* **54** 277-305

[8] T Nakano, N Ohnuki, S Baba 2000 Pressure dependence of optical emission from DC magnetron sputtering plasma observed with spatial resolution *Vacuum* **59** 581-585

[9] N Nafarizal, N Takada and K Sasaki 2009 Investigations of production processes of Ti+ in high-pressure magnetron sputtering plasmas *Jpn. J. Appl. Phys.* **48** 126003

[10] S R How, N Nayan, J Lias 2017 The influence of N2 flow rate on Ar and Ti Emission in high-pressure magnetron sputtering system plasma *AIP Conference Proceedings* **1824** 030024

[11] NIST Atomic Spectra Database http://www.nist.gov/pml/data/asd.cfm, (Access date: 2.2.2017)

[12] M M Yusoff, M H Mamat, M F Malek, A B Suriani, A Mohamed, M K Ahmad, Salman A H Alrokayan, Haseeb A Khan and M Rusop 2016 Growth of titanium dioxide nanorod arrays through the aqueous chemical route under a novel and facile low-cost method *Materials Letters* **164** 294-298

[13] E Bultinck, S Mahieu, D Depla and A Bogaerts 2009 Reactive sputter deposition of TiNx films, simulated with a particle-in-cell/Monte Carlo collisions model *New Journal of Physics* **11** 023039

IOP Conf. Series: Journal of Physics: Conf. Series **1027** (2018) 012005 doi:10.1088/1742-6596/1027/1/012005

[14] K Tao, D Mao and J Hopwood 2002 Ionized physical vapor deposition of titanium nitride: A global plasma model *J. Appl. Phys.* **91** 4040-4048
[15] S R How, N Nayan, J Lias 2017 A simple temperature evaluation in high-pressure magnetron sputtering plasma using optical emission spectroscopy (OES) technique *AIP Conference Proceedings* **1788** 030054
[16] N Britun, M Gaillard, A Ricard, Y M. Kim, K S Kim and J G Han 2007 Determination of the vibrational, rotational and electron temperatures in N2 and Ar-N2 rf discharge *J. Phys. D: Appl. Phys.* **40** 1022-1029

IOP Conf. Series: Journal of Physics: Conf. Series **1027** (2018) 012006 doi:10.1088/1742-6596/1027/1/012006

Effects of photocatalytic activity of metal and non-metal doped Tio_2 for Hydrogen production enhancement - A Review

Siti Nur Aqilah Sulaiman[1], Mohamad Zaky Noh[1], Nurul Nadia Adnan[1], Noriah Bidin[2], Siti Noraiza Ab Razak[*3]

[1]Department of Physics and Chemistry, Faculty of Applied Science and Technology, Universiti Tun Hussein Onn Malaysia, 84600, Pagoh, Johor, Malaysia
[2]Laser Center, Universiti Teknologi Malaysia, 81310, UTM, Skudai, Johor, Malaysia
[3]Department of Science and Mathematics, Centre for Diploma Studies, Universiti Tun Hussein Onn Malaysia, 84600, Pagoh, Johor, Malaysia

Corresponding author: noraiza@uthm.edu.my

Abstract. Titanium dioxide TiO_2 is well-known materials that has become an efficient photocatalyst for environmental sustainability. Known as solar driven catalysis, TiO_2 is considered as the most promising way to alleviate environmental issues caused by the combustion of fossil fuels and to meet worldwide demands for energy. Much effort has been concerned on TiO_2 band gap modification to become a visible-light-activated photocatalysts of TiO_2 because it can only be excited by UV light irradiation due to its large band gap. Modifications like metals and nonmetals doping has been proposed in the past decades. This reviews survey recent advanced preparation methods of doped-TiO_2 including various types of doping methods for various types of dopants and provides general review on further modifications. The characterizations techniques used in order to determine the structural, morphological and optical properties of modified TiO_2 is also discussed. Further, a new method of TiO_2 modification is proposed in this mini review paper.

1. Introduction

Titanium dioxide (TiO2) or commonly known as titania is the most effective photocatalyst for air and water purification and self-cleaning surfaces, and also widely used for energy-related fields especially in water splitting under visible light irradiation [1-3]. According to previous research, TiO2 has been proven to be the most efficient and economic photocatalyst due to its capability to completely mineralize contaminants [4-5], low cost material, superhydrophilicity [6,8] for self-cleaning function and antibacterial agent [9], non-toxicity, and has high oxidative abilities [10-12].

Photocatalysis was first discovered by Fujishima in 1970's and later called as Honda-Fujishima Effect. A strong oxidation and reduction power of TiO2 was realized through a simple experiment of TiO2 electrode in an aqueous solution exposed to strong light. Photoinduced decomposition of water on TiO2 electrodes are reported as a gas bubbles evolved from the surface of electrodes after being exposed to strong light. Analogically, photocatalytic reaction on TiO2 was found to be similar with photosynthesis – as illustrated in Figure 1; light is used as an energy that enables to break water molecule into hydrogen and oxygen [13].

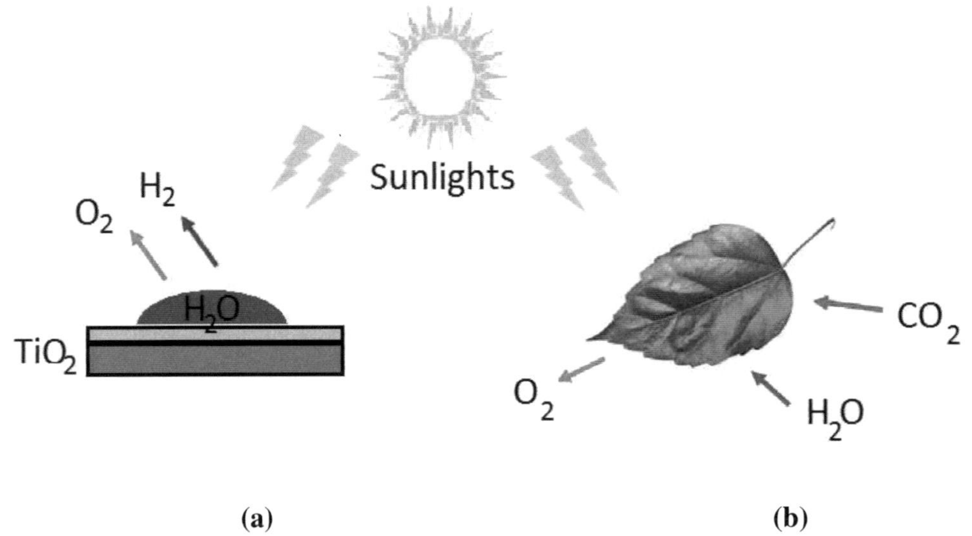

Figure 1. Light is absorbed by (a) titania to break water molecule into hydrogen and oxygen, and (b) chlorophyll during photosynthesis to produce oxygen gas [13].

Following from Honda-Fujishima effect, the applications of TiO_2 photocatalyst are elevates rapidly in various sector such as in photoelectrolysis to generate hydrogen energy from water splitting [14-21], anti-corrosion coating material [22], and an organic pollutant degradation to tackle environmental issues [23-27]. In general, when TiO_2 surface was irradiated with ultraviolet light there will be two types of photochemical reaction to occur. First, photo-induced redox reactions of adsorbed substances, and also photo-induced hydrophilic conversion of TiO_2 itself.

Despite all the interesting features in TiO_2, its large energy band gap is in range 3.0 eV (rutile) to 3.5 eV (anatase) corresponding to wavelengths of 385 and 410 nm respectively – which allows the absorption of light in ultraviolet range only [28]. Besides, fast recombination of electrons and holes, slow charge carrier transfer and also requires high reproduction cost limit the efficiency of photocatalytic activity of TiO_2 [29]. Due to this flaw, various attempts has been done in order to modified the band gap of TiO_2; with such it will broaden the light absorption range mainly in visible light range that corresponds about 40 % of light spectrum [28,30-33]. The photocatalytic activity under simulated sunlight irradiation is determine from the band gap, the valence band VB and covalence band CB edge positions of photocatalyst [17].

Many approach have been done in previous research to reduce the intrinsic band gap of titania including cations (mainly transition metals or rare earth metal ions) or anions doping into TiO_2 lattice host material [12,28,34-38].

2. Preparation of Doped TiO_2

There are various chemical and physical methods available on preparing doped-TiO_2 - either anatase or rutile structure. Those methods including ion-assisted sputtering, plasma, ion-implantation, chemical vapor deposition (CVD) and sol-gel to enhance the photocatalytic effect in the visible light region. A list of doped TiO_2 and methods of its preparation are shown in Table 1(a) and Table 1(b) [10,39-42].

2.1 TiO_2 Doped with Metal Species

Metal doped TiO2 especially transition metal has been extensively investigated over the past decades for enhancing its photocatalytic performance on the degradation of various organic pollutants under visible irradiation. This modification is done to extend the optical absorption of TiO2 to visible light.

Table 1(a). Dopants and Preparation Methods of metal-doped-TiO_2 photocatalysts.

Doped Element	Preparation Method	Reference
Pd	Synthesized using alkaline route of sol-gel method. The doped materials were obtained by simultaneous gelation of both precursors in the sol–gel process. Absolute ethanol, min. 99.8 vol %, was used as solvent for both titania and palladium precursors. The hydrolysis reaction took place in the presence of the water, added in substoichiometric quantity, in a closed system, in nitrogen atmosphere, under vigorous stirring. The nitric acid, HNO_3 65 % Merck, was used as catalyst to obtain a final pH of the sol in the range 3-3.5.	[39]
Fe	Titanium tetraisopropoxide, 97% pure, ethanol, diethanolamine and iron nitrate, 98% pure, were used for preparation of thin films. Indium doped tin oxide (SnO_2:In) conducting glass was used as substrate. A transparent gel solution of doped/ undoped TiO_2 was prepared by mixing of 3 ml TTIP and calculated amount of iron nitrate in 20 ml ethanol in the presence of diethanolamine. The solution was stirred for 4 hours at room temperature to enhance the reaction rate between diethanolamine and TTIP and finally it gets converted into gel. This gel solution was applied on nearly two third area of the conducting glass substrate and uniformly coated with spin coating unit. After each successive layer the film was allowed to dry for 10 min at 80 °C. Final sintering was done at 500 °C for 2 h.	[40]

It is reported that metal doping has shown photocatalytic activities at both UV and visible light range [43]. Furthermore, metal-doped TiO_2 will caused many drawbacks including dopant insolubility, thermal instability during the process, phase transformation among the titania polymorphs, the increase of carrier-recombination centers, and the alteration of charge carrier diffusion length.

2.2 TiO_2 With Non-metal Species

Nonmetal-doped TiO_2 is reported as the most promising photocatalysts candidates [44]. An extensive research of nonmetal-doped TiO_2 photocatalysts was widely carried out since the early 1990s [5,8]. Among all the nonmetal-doped TiO_2 photocatalysts, nitrogen-doped TiO_2 is the most studied material [45]. In addition, several other visible-light active materials, such as carbon C, F and S, doped TiO2 were reported [19].

IOP Conf. Series: Journal of Physics: Conf. Series **1027** (2018) 012006 doi:10.1088/1742-6596/1027/1/012006

Table 1(b). Dopants and Preparation Methods of non-metal-doped-TiO_2 photocatalysts.

Doped Element	Preparation Method	Reference
Glass Substrate	The three components were used in the preparation of the TiO_2 Sol-Gel films it can be explained as follows: Titanium Isopropoxide (TIP), Acetic acid and Ethanol. TiO_2 thin films are prepared by Sol-Gel spin- coating method using 50 ml of Ethanol and allowed to mix for 50 min, measure 5 ml of acetic acid with a pipette into Ethanol and stir by using a magnetic stirrer it for 5 min or at least 3 min, and measure 6.3 ml of TIP was added by pipette to a beaker containing a mixture of glacial acetic acid and ethanol that had been mixed for five minutes. The mixture was continually stirred using a magnetic stirrer addition and for a further two minutes after addition of the precursor. The concentration was altered in order to assess the effects of ratios and concentration. Glass slide coated with transparent conducting indium tin oxide (ITO) films were used as substrates.	[10]
Zz N, S, C	Doping was performed by an incipient wet impregnation method as the N, C and S-sources. TiO_2 Degussa P25 and appropriate amounts of N, C and S were mixed with specific volumes of doubly distilled water and stirred for 1 hour. During this period, the mixture changed color into a yellowish/ greyish white depending upon the anion concentration. Then, the prepared photocatalysts were washed with water three times, heat-treated at 373 K for 24 hours to eliminate water, calcined at three different temperatures in the range 623-973 K for 1, 3 and 5 hours, ground and sieved.	[41]
N	TiO_2 nanostructures were prepared using LPD method. ITO substrates underwent standard cleaning procedure where they were cleaned using acetone and ethanol in sequence. The growth process was carried out under room temperature. The as prepared nanostructured TiO_2 was rinsed and dried before it was annealed at 400 °C for 30 minutes. Growth of the TiO_2 nanoparticles was started with hydrolysis of the metal flouro cation used ($(NH_4)_2TiF_6$) with boric acid was used to scavenge the flouro cation left. TiO_2 nanoparticles film formed on the surface was then subjected towards annealing process to purify the metal oxide nanoparticles produce and to remove organic materials left during the hydrolysis process. The treated samples were then subjected towards heating at 100 °C for 15 minutes to remove any hydroxyl compounds.	[42]

The three components were used in the preparation of the TiO_2 Sol-Gel films it can be explained as follows in Figure 2.

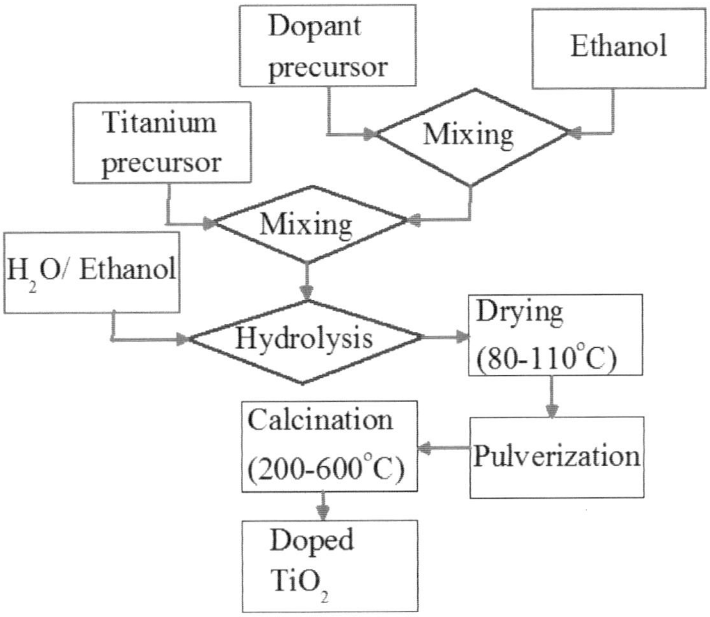

Figure 2. Procedure of doped-TiO_2 via Sol-Gel method.

TiO_2 thin films are prepared by sol-gel spin- coating method using 50 ml of ethanol and allowed to mix for 50 min, measure 5 ml of acetic acid with a pipette into ethanol and stir by using a magnetic stirrer it for 3 to 5 min. The mixture was continually stirred using a magnetic stirrer addition and for a further two minutes after addition of the precursor. The sol-gel concentration was altered in order to assess the effects of ratios and concentration. The effects of altering molar ratios of acid are investigated. The Sol-Gel manufactured due to the method described above was immediately used to produce coating on the various substrates. Thin films of TiO_2 were deposited by spin coating method using a standard photoresist spinner [1,10,46].

In short, the Sol-Gel process is one of the versatile methods in preparing Nano-size materials. This technique does not require a complicated instrument which provides a simple and easy means for preparing Nano-size particles. The incorporation of an active dopant in the Sol-Gel during the gelation stage allows the doping elements to have a direct interaction with support, therefore, the material possesses catalytic or photocatalytic properties. Titanium precursor, such as titanium isoprooxide (TIP), tetrabutyl orthotitanate (TBOT), titanium tetra chloride, is mixed with dopant precursor dissolved in alcohol, followed by hydrolysis performed at the room or elevated temperature. The precipitate is dried usually at temperature range from 80 to 110°C, pulverized to obtain xerogel and calcinated in air at temperature from 200 to 600°C [46].

3. Morphological Surface and Photocatalytic Analysis

Among metal dopants as mentioned previously, Fe (III) ions as the most favourable dopant due to its unique electronic structure and its similar size with titanium (IV), Ti^{4+} ions [40]. The energy level of the Fe^{4+}/Fe^{3+} couple is only just greater than the titania conduction band and the energy level of the Fe^{3+}/ Fe^{2+} couple is just above the valance band [47]. This useful electronic states of iron ions in titania results in formation of efficient trapping sites for electrons and holes [48].

3.1 Crystalline structure of doped-TiO2

Aside of types of dopant substitute into TiO2 lattice structure, the annealing temperature of TiO_2 thin films also plays a crucial roles in order to produce a rutile or anatase semiconductor. A pure TiO_2 films is prepared at various annealing temperature from 300 to 800 °C. Figure 3 shows the X-ray diffraction XRD patterns of the prepared samples. It was found that the pure TiO_2 films with anatase phase increase when annealing temperature increased. The pure TiO_2 films formed of anatase phase show a very high efficiency photocatalytic activity due to their large internal surface. Both the optical properties and the photocatalytic activity of TiO_2 coatings depend strongly on the crystalline phase, the crystallite size and the porosity of the coatings. Moreover, the TiO_2 films formed of rutile phase when the sample annealed at a temperature more than 700 °C [6].

Figure 4 shows the XRD pattern of pure TiO_2 and 0.2 at.% Fe doped TiO_2 films at final sintering temperature at 500 °C for 2 h. . The obtained patterns shows a predominance of sharp, intense peak corresponding to underlying SnO2:In layer on the substrate. From the pattern, weak intensity peaks corresponding to anatase phase of TiO_2 is obtained for all samples of Fe^{3+} doped TiO_2. Besides, a high crystalline material was obtained for 0.2 at.% Fe doped TiO_2. By using Scherrer's equation ($D = K\lambda/\beta\cos\theta$), the crystallite size was calculated and the sizes of TiO_2 particles are confirmed in nanosize for different concentration of dopant [40].

As for non-metal doped TiO_2, a carbon-doped TiO_2 thin films were prepared by hydrothermally synthesized using $Ti(SO_4)_2$ and glucose as titania and carbon resources for 12 to 48 hours [19]. The obtained carbon-TiO_2 nanocrystals display higher energy conversion efficiencies than pure TiO_2 when applied as the photo-anode materials in dye-sensitized solar cells DSSCs. Based on the combinational analyses on dye adsorption efficiency, photoluminescence and electron lifetime of TiO_2, the retarded hole-electron recombination rate and the negatively shifted flat band potential were revealed as the main reasons for the improved energy conversion efficiency.

XRD pattern in Figure 5 shows an obvious anatase phase peaks in each sample. This proves that the presence of glucose does not cause the change of crystalline phase. The calculated crystallite size of carbon-TiO_2 is in the range of 14 to 19 nm that is smaller than that of pure TiO_2 (24 nm). Smaller size of TiO_2 nanoparticles will improved particle dispersity for conversion efficiency advancement.

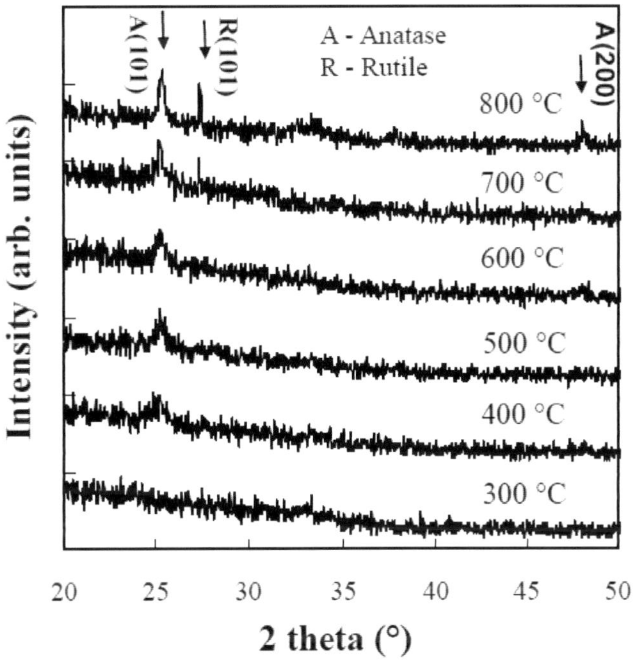

Figure 3. X-ray diffraction patterns of pure TiO_2 films after annealed at different temperature [6].

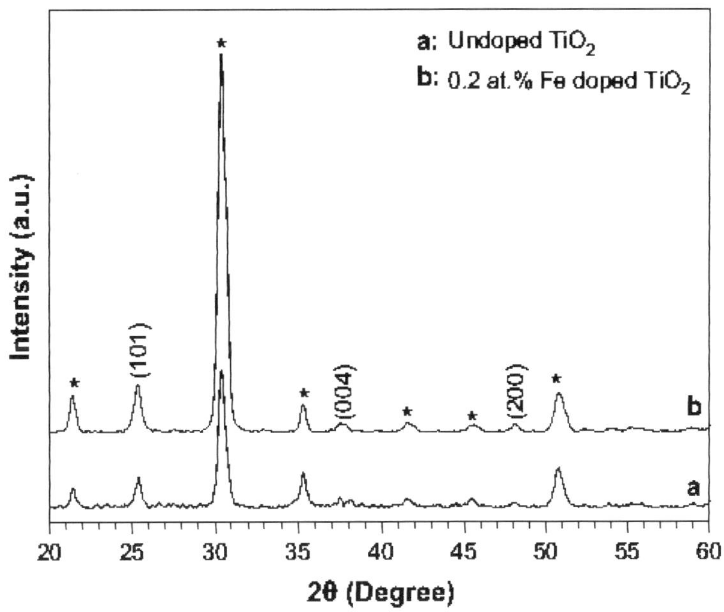

Figure 4. XRD patterns of pure TiO_2 (undoped) and doped 0.2 at.% Fe thin films deposited on conducting glass plate [40].

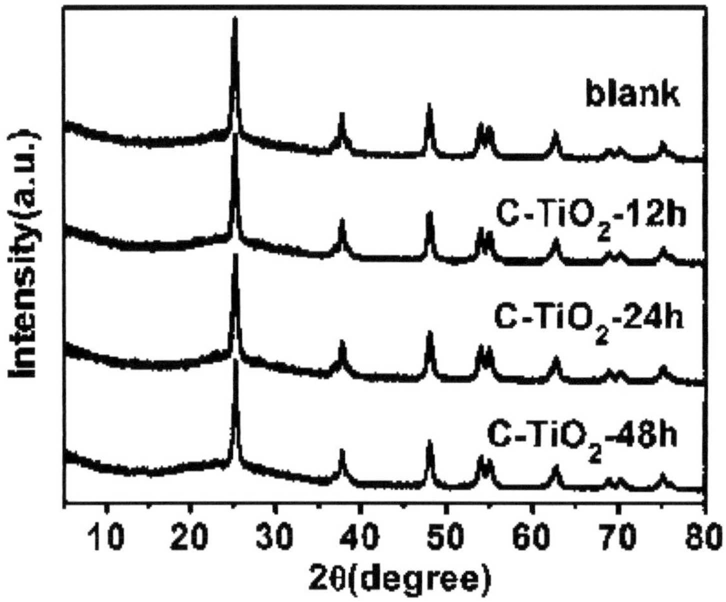

Figure 5. X-ray diffraction patterns of pure TiO_2 carbon-doped TiO_2 films [19].

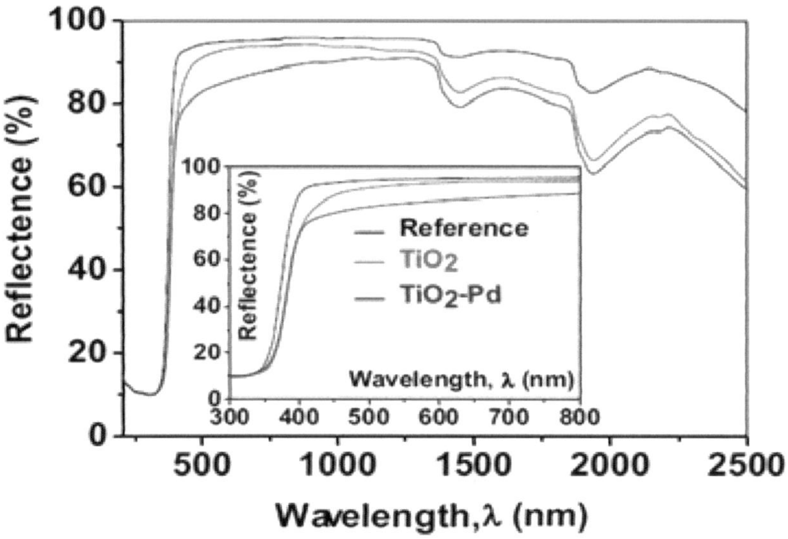

Figure 6. UV–vis–NIR diffuse reflectance spectrophotometric curves for pure and Pd-doped-TiO_2 nanoparticles [52].

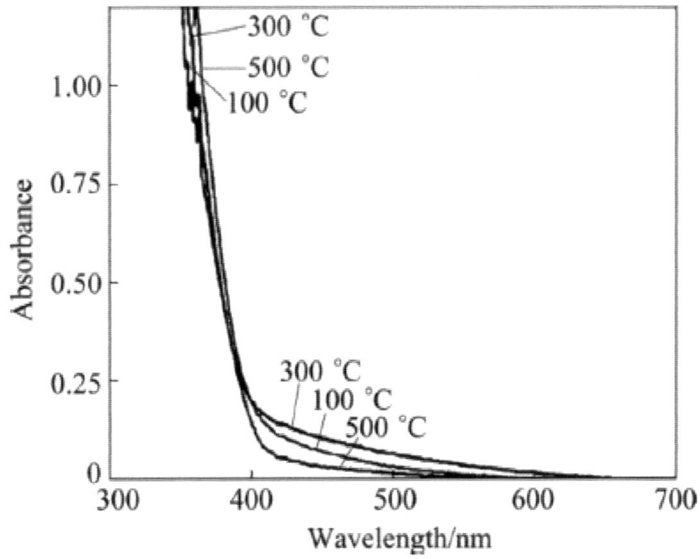

Figure 7. UV-vis absorption curves of Zn-doped-TiO2 films [53].

The band gap of the TiO_2:Zn^{2+} is decreased with increasing Zn^{2+} dopant. Since the valence band edge of TiO_2 material is dominated by O 2p, and the conduction band edge is formed from Ti 3d, some of the new occupied molecular orbitals located below the conduction band of TiO_2 would be formed when Zn^{2+} ions were doped into TiO_2 lattice. The decreased of band gap would likely cause to the charge transfer from the dopant energy level Zn^{2+} to the conduction band of TiO_2 [47]. Besides, other dopant that shows an interesting result is carbon. C-doped-TiO_2 proved that the particles are crystalline with spherical morphology, high surface area and an optical band gap of 2.79 eV for C-TiO_2 annealed at 400 °C as reported by Warkhade et. al. [11].

4. Conclusion

The modification methods of *x*-doped-TiO_2 have been discussed in this review paper. Between two types of dopant elements that are metal and non-metal, non-metal (anions) shows a significance role in order to enhance the performance of TiO_2 photocatalyst. The biggest challenge involving TiO2 as photocatalysts is to increase its spectral sensitivity to visible light. A new dopants and new method in preparing modified structure of TiO_2 would be the next approach of future research. With such, a new applications of TiO_2 in various sector would be revealed. Currently, the major problem with doped TiO_2 may be the loss of photoactivity during recycling and long-term storage. It is seems that the efficiency of doped-TiO_2 under visible light strongly depended on the preparation method used.

Acknowledgement

The author would like to express their thanks to government Malaysia for the financial support in this project through FRGS grant 1620, and also UTHM through RMC for their management of the grant.

References

[1] R Dholam, N Patel, M Adami and A Miotello 2008 Physically and chemically synthesized TiO 2 composite thin films for hydrogen production by photocatalytic water splitting *Int. J. Hydrogen Energy* **33**(23) 6896–6903

[2] M Ni, M khla, D Y C Leung and K Sumathy 2007 A review and recent developments in photocatalytic water-splitting using TiO 2 for hydrogen production **11** 401–425

[3] Z Zhu, C Kao, B Tang, W Chang and R Wu 2016 Efficient hydrogen production by photocatalytic water-splitting using Pt-doped TiO 2 hollow spheres under visible light *Ceram. Int.* **42**(6) 6749–6754

[4] R J Tayade, S Technol and A Mater 2007 Photocatalytic degradation of dyes and organic contaminants in water using nanocrystalline anatase and rutile TiO 2 Photocatalytic degradation of dyes and organic contaminants in water using nanocrystalline anatase and rutile TiO 2

[5] C S Turchi 1990 Photocatalytic Degradation of Organic Water Contaminants : Mechanisms Involving Hydroxyl Radical Attack **192** 178–192

[6] S Lien and J Jhu applications 2–4

[7] H Kong 2014 Light-induced super-hydrophilicity and photocatalytic activity of mesoporous TiO2 thin films no. May 2002

[8] M Machida, K Norimoto and T Watanabe 1999 The effect of SiO 2 addition in super-hydrophilic property of TiO 2 photocatalyst **4**, 2569–2570

[9] A Zaleska 2008 Doped-TiO 2 : A Review (1) 157–164

[10] V S Process, M H Shinen, K I Ajeel and F A Rasin 2001 Preparation of Titanium Dioxide (TiO 2) no. Rancourt 1987 1–9

[11] S K Warkhade, G S Gaikwad, S P Zodape, U Pratap and A V Maldhure 2017 Low temperature synthesis of pure anatase carbon doped titanium dioxide : An efficient visible light active photocatalyst *Mater. Sci. Semicond. Process.* **63**(November 2016) 18–24

[12] A O Technologies 2014 The Role of Non-Metal Doping in TiO2 Photocatalysis The Role of Non-Metal Doping in TiO 2 Photocatalysis no. April

[13] K Hashimoto, H Irie and A Fujishima 2005 TiO 2 Photocatalysis: A Historical Overview and Future Prospects *Jpn. J. Appl. Phys.* **44**(12) 8269–8285

[14] B Gupta and A A Melvin 2017 TiO 2 / RGO composites : Its achievement and factors involved in hydrogen production *Renew. Sustain. Energy Rev.* **76** April 1384–1392

[15] D Y C Leung, X Fu C Wang, M Ni and M K H Leung 2010 Hydrogen Production over Titania - Based Photocatalysts Hydrogen Production over Titania-Based Photocatalysts no. April 2014

[16] S Escobedo, B Serrano, A Calzada, J Moreira and H De Lasa 2016 Hydrogen production using a platinum modified TiO 2 photocatalyst and an organic scavenger Kinetic modeling *Fuel* **181** 438–449

[17] F Opoku, P Govender, F Opoku, K K Govender, C Gertina and C Elizabeth 2017 Recent Progress in the Development of Semiconductor-Based Photocatalyst Materials for Applications in Recent Progress in the Development of Semiconductor- Based Photocatalyst Materials for Applications in Photocatalytic Water Splitting and Degradation no. July

[18] T U Wien 2016 Hydrogen production by photocatalytic water- splitting on Pt-doped TiO2 – ZnO under visible light *J. Taiwan Inst. Chem. Eng.* no. November

[19] A Lin, D Qi, H Ding, L Wang, M Xing, B Shen and J Zhang 2017 Carbon-doped titanum dioxide nanocrystals for highly efficient dye-sensitized solar cells *Catal. Today* **281** 636–641

[20] P Wang, S Zhan, Y Xia, S Ma, Q Zhou and Y Li 2017 Applied Catalysis B : Environmental The fundamental role and mechanism of reduced graphene oxide in rGO / Pt-TiO 2 nanocomposite for high-performance photocatalytic water splitting **207** 335–346

[21] F Song 2017 Metal – Organic frameworks and their derivatives for photocatalytic water

splitting inorganics Metal – Organic frameworks and their derivatives for photocatalytic water splitting no. July

[22] K Nakata and A Fujishima 2012 Reviews TiO 2 photocatalysis : Design and applications *Journal of Photochemistry and Photobiology C : Photochemistry* **13** 169–189

[23] W Kim, G Jang, J Lee and D Rhee 2014 Degradation of Methylene Blue by Titania doped with Transition Metal and Nitrogen *Energy Procedia* **61** 2456–2459

[24] M Reza, D Khaki, M Saleh, A Aziz, A Raman, W Mohd and A Wan 2017 Application of doped photocatalysts for organic pollutant degradation - A review *J. Environ. Manage.* **198** 78–94

[25] S Bagheri and N M Julkapli 2016 ScienceDirect Synergistic effects on hydrogenated TiO 2 for photodegradation of synthetic compounds pollutants *Int. J. Hydrogen Energy* **41**(33) 14652–14664

[26] P Wen, M Hayrie, M Hatta, S Teng, L Yuliati and S Ling 2017 Journal of Photochemistry and Photobiology A : Chemistry Photocatalytic degradation of photosensitizing and non-photosensitizing dyes over chromium doped titania photocatalysts under visible light *Journal Photochem. Photobiol. A Chem.* **332** 215–223

[27] B Appavu and S Thiripuranthagan 2017 Journal of Photochemistry and Photobiology A : Chemistry Visible active N , S co-doped TiO 2 / graphene photocatalysts for the degradation of hazardous dyes *Journal Photochem. Photobiol. A Chem.* **340** 146–156

[28] S Z Islam, S Nagpure, D Y Kim and S E Rankin 2017 Synthesis and Catalytic Applications of Non- Metal Doped Mesoporous Titania Synthesis and Catalytic Applications of Non-Metal Doped Mesoporous Titania *Inorganics* **5**(1) 15

[29] M Ge, J Cai, J Iocozzia, C Cao, S Zhang, K Zhang, Y Lai and Z Lin 2016 Science Direct A review of TiO 2 nanostructured catalysts for sustainable H 2 generation *Int. J. Hydrogen Energy* **42**(12) 8418–8449

[30] H Yan, X Wang, M Yao and X Yao 2013 Band structure design of semiconductors for enhanced photocatalytic activity: The case of TiO2 *Prog. Nat. Sci. Mater. Int.* **23**(4) 402–407

[31] H Ceylan, C Ozgit-Akgun, T S Erkal, I Donmez, R Garifullin, A B Tekinay, H Usta, N Biyikli and M O Guler 2013 Size-controlled conformal nanofabrication of biotemplated three-dimensional TiO_2 and ZnO nanonetworks *Sci. Rep.* **3** 2306

[32] A V Rosario and E C Pereira 2014 Applied Catalysis B : Environmental The role of Pt addition on the photocatalytic activity of TiO 2 nanoparticles : The limit between doping and metallization *Applied Catal. B, Environ.* **144** 840–845

[33] O Van Overschelde, R Snyders and M Wautelet 2007 Crystallisation of TiO2 thin films induced by excimer laser irradiation *Appl. Surf. Sci.*, **254**(4) 971–974

[34] L Lei, Y Su, M Zhou, X Zhang and X Chen 2007 Fabrication of multi-non-metal-doped TiO2 nanotubes by anodization in mixed acid electrolyte *Mater. Res. Bull.* **42**(12) 2230–2236

[35] K Siuzdak, M Abbas, L Vignau, M Devynck, G V Dubacheva and A Lisowska-Oleksiak 2012 Application of non-metal doped titania for inverted polymer solar cells *J. Appl. Phys.* **112**(12)

[36] C Di Valentin and G Pacchioni 2013 Trends in non-metal doping of anatase TiO 2 : B , C , N and F **206** 12–18

[37] K Kadakia, M K Datta, O I Velikokhatnyi, P Jampani, S K Park, S J Chung and P N Kumta 2014 High performance fluorine doped (Sn,Ru)O2 oxygen evolution reaction electro-catalysts for proton exchange membrane based water electrolysis *J. Power Sources* **245** 362–370

[38] O Linnik, N Shestopal, N Smirnova, A Eremenko, O Korduban, V Kandyba, T Kryshchuk, G Socol, N Stefan, G Popescu-pelin, C Ristoscu and I N Mihailescu 2015 Correlation between electronic structure and photocatalytic properties of non-metal doped TiO 2 / ZrO 2 thin films obtained by pulsed laser deposition method *Vaccum* **114** 166–171

[39] M Cris, A Bra, D Cris, N Dra, A Ianculescu, D Mardare, D Luca, V Marinescu and A Moldovan 2008 Crystallization study of sol – gel un-doped and Pd-doped TiO 2 materials *Journal of Physics and Chemistry of Solids* **69** 2548–2554

[40] A P Singh, S Kumari, R Shrivastav, S Dass and V R Satsangi 2008 Iron doped nanostructured TiO 2 for photoelectrochemical generation of hydrogen *Int. J. Hydrogen Energy* **33**(20) 5363–5368

[41] Y Yalçin, M Kiliç and Z Çinar 2010 The role of non-metal doping in TiO 2 photocatalysis *J. Adv. Oxid. Technol.* **13**(3) 281–296

[42] A A Umar, M Yusri, A Rahman, S Khatijah, M Saad and M M Salleh 2012 Effect of NH 3 Concentration on the Performance of Nitrogen doped TiO 2 Photoelectrochemical Cell *Int. J. Electrochem. Sci* **7** 7855–7865

[43] J Park, K Choi, J Lee, C Hwang, D Choi and J Lee 2013 Fabrication and characterization of metal-doped TiO 2 nanofibers for photocatalytic reactions *Mater. Lett.*, **97** 64–66

[44] T Teka 2015 Current State Of Doped-Tio 2 Photocatalysts And Synthesis Methods To Prepare Tio 2 Films : A Review **3**(1) 14–18

[45] S Walck 2014 Photochemical Activity of Nitrogen-Doped Rutile TiO2 (110) in Visible Light **2**

[46] A Zaleska 2008 Doped-TiO2: A Review *Recent Patents Eng.*, **2**(3) 157–164

[47] T T Loan, V H Huong, V T Tham and N N Long 2017 Effect of zinc doping on the bandgap and photoluminescence of Zn 2 + -doped TiO 2 nanowires 2–7

[48] A Zarubica, T Minovi and N Abazovi 2016 Journal of the European Ceramic Society Iron doped anatase for application in photocatalysis **36** 2991–2996

[49] S Chen 2014 Band gap narrowing of TiO2 doped with Ce probed with X-ray absorption spectroscopy 2–5

[50] T Morikawa and T Central 2001 BandGap Narrowing of Titanium Dioxide by Nitrogen Doping Japanese Journal of Applied Physics no. May 2016 2–5

[51] C O Ayieko, R J Musembi, S M Waita, B O Aduda and P K Jain 2012 Structural and Optical Characterization of Nitrogen-doped TiO 2 Thin Films Deposited by Spray Pyrolysis on Fluorine Doped Tin Oxide (FTO) Coated Glass Slides **2**(3) 67–72

[52] A Narayan, N Hamnabard and S Woo 2016 A comparative study of the effect of Pd-doping on the structural, optical and photocatalytic properties of sol – gel derived anatase TiO 2 nanoparticles *Ceram. Int.*, **42**(10) 12010–12026

[53] L Qiao, F Xie, M Xie, C Gong, W Wang and J Gao 2016 Characterization and photoelectrochemical performance of Zn-doped TiO 2 films by sol – gel method *Trans. Nonferrous Met. Soc. China* **26**(8) 2109–2116

Optical fiber sensor based on a polymer optical fiber macro-bend to study thermal expansion of metals

Paradorn Pakdeevanich[1]

[1]Department of Physics, Faculty of Science, Silpakorn University, Nakorn Prathom, Thailand

Corresponding author: pakdeevanich_p@su.ac.th

Abstract. Thermal expansion is an important parameter for characterization of metals. As metal is heated, the molecules vibrate more violently and expand in all direction. Investigators have focused to study the thermal strain. However, the amount of expansion is difficult to measure. An attempt has been made to develop an apparatus using optical technique. The principle of this system is the transformation of length changes into changes of light intensity. The purpose of this work is to design and develop an optical fiber sensor based on a macro-bend of a polymer optical fiber. In this system, thermal expansion of metal was converted into the rolling of a needle in which placed beneath a flat bar of metal. Optical fiber sensor was attached to the ended section of a needle. As the crimp tube of the fiber sensor was moved due to thermal expansion of metal, the bend radii of optical fiber sensor was changed. As a sequence, the loss induced by the bending effect was depended on the expansion of metal that changed with temperature. In this study, we utilized optical fiber sensor to monitor and compare the thermal expansion of copper, brass and aluminum. According to our experimental results, the linear response with temperature was reported. The measured values of coefficient of thermal expansion was analyzed to be 0.45 , 0.35 and 0.32 a.u./°C for aluminum bar, brass bar and copper bar, respectively. In addition, the effect of the size of the diameter of a needle on the response of bending loss was investigated.

1. Introduction

Thermal expansion is a fundamental property of material. It relates to the dimensional changes of material when temperature is changed. Fundamentally, for small change in temperature (ΔT), the length change (ΔL) is proportional to the temperature change. A convenient measure of thermal expansion is the linear coefficient of thermal expansion (α). The linear coefficient of thermal expansion is defined as [1]

$$\alpha = \frac{\Delta L}{L_0[\Delta T]} \tag{1}$$

where L_0 is the original length of the object. In other word, α measures the amount of strain which accompany a material with a change in temperature. Accurate measurement of α is very important because it reflects the binding force in solid. Previously, S. Kanagaraj and S. Pattanayak [2] reviewed comprehensively of the various techniques to explore the thermal expansion of metals. Interestingly, there were various research groups made effort to conduct optical technique for investigation of thermal expansion in solid. We found that interferometric principle was widely developed. Wolff and Eselun [3] modified Michelson interferometer to study thermal expansion of fused silica and quartz tube. However, all mentioned techniques required the complex set-up equipment and the systems were gigantic. Therefore, we decided to explore for the optical fiber system to cope with this task.

In the last decade, sensors based on fiber optics have demonstrated very good performance to deal with various monitoring. Also, it is simplicity, low cost and immune to electromagnetic interference. Due to those advantages, we had a strong focus on the creation of optical fiber sensor to provide an effective measurement which could compete with the conventional method. A number of useful reviews of fiber optic sensor technology had been discussed in detail by K.T.V. Grattan and B.T. Meggitt [4].

Basically, a light ray absolutely reflected at core-cladding interface if the incidence angle (θ) is greater than the critical angle (θ_c) as described in Figure 1(a). When its angle of incidence is inferior to the critical angle $(\theta < \theta_c)$, at these corners the light will not satisfy the condition for total internal reflection and hence the refracted wave escapes out from the fiber as described in Figure1(b). Therefore, an optical fiber that one bending beyond a critical radius causes optical power guided mode experiences loss as it goes through a bent section and it escapes the core. This is called as macroscopic bending loss [5].

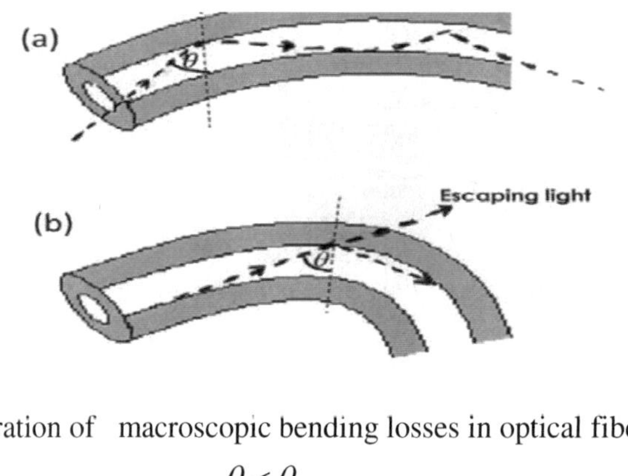

Figure 1. Geometrical illustration of macroscopic bending losses in optical fiber: (a) $\theta > \theta_c$ and (b) $\theta < \theta_c$.

In short, we proposed an invention of fiber optic sensor which based on fiber bending loss. In recent years, applications of bending loss to innovate sensor have been reported by the following research groups. Liu et al [6] innovated displacement sensor based on the macro-bend coupling effect which caused power transmission between two twisted bending plastic optical fibers. Gattass et al [7] demonstrated the bending loss for multimode chalcogenide fiber over a wide range of infrared wavelength. They concluded that the bending loss was not dependent on the wavelength used. Obviously, their measured results indicated that there was increasing loss significantly due to tight bends. The design and development of a polymer optical fiber (POF) was reported to be used for temperature sensing[8]. It based on the concept that macrobending loss caused by the different thermo-optic coefficient of the cladding and core of POF. At this point it was challenged us to design and develop a simple sensor which required only a relative low-cost component, compact size and easy to build. Crucially, our innovated sensor had an intuitive capability of detecting at quite small change of thermal expansion.

2. Experimental setup and procedure

The design system is presented in Figure 2. First, the metal bar was grabbed to fix by one side. By direct contact to the hot plate, heat transferred by conduction caused metal to expand on the free side. Since the expansion was quite small, it needed a trick to amplified. At the free side of metal, a needle was placed beneath a bar of metal. Basically, thermal expansion of metal was converted into the rolling of a needle. Thus, rolling of needle was a key monitoring. As mentioned before, we intended to propose optical fiber sensor. In this situation, we planned to convert the amount of this rolling to relate with light intensity of optical fiber sensor. In order to approach the required intention, the optical fiber sensor was attached to one end of a needle.

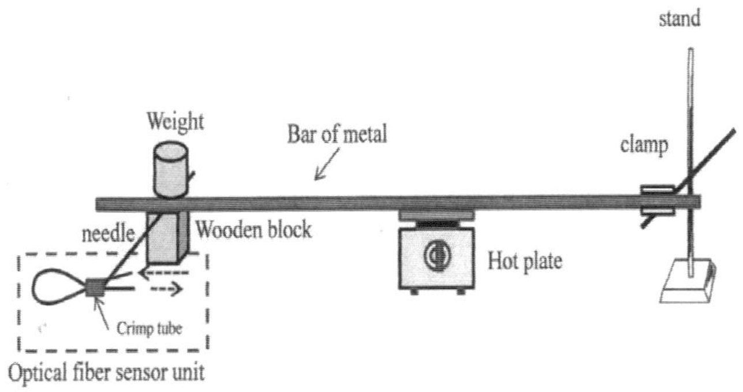

Figure 2. Schematic diagram of the system to convert the thermal expansion into the rolling of a needle.

The whole optical fiber sensor unit was illustrated in Figure 3. The setup system was composed of He-Ne Laser, the macrobending loss sensor and photodetector.

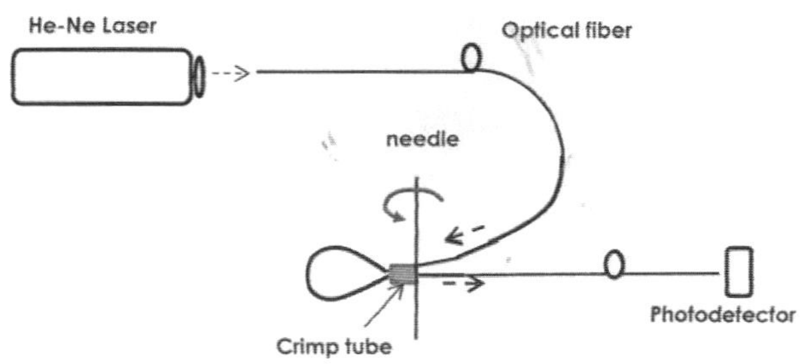

Figure 3. Schematic diagram of the optical fiber sensor unit.

Figure 4 was attempted to explain the mechanism of this sensor. As the crimp tube of the fiber sensor was moved (via the rolling of needle) due to thermal expansion of metal, the bend radius of optical fiber sensor was changed as compared in Figure 4(a) and 4(b). Then, the loss induced by the bending effect was depended on the expansion of metal that changed with temperature.

IOP Conf. Series: Journal of Physics: Conf. Series **1027** (2018) 012007 doi:10.1088/1742-6596/1027/1/012007

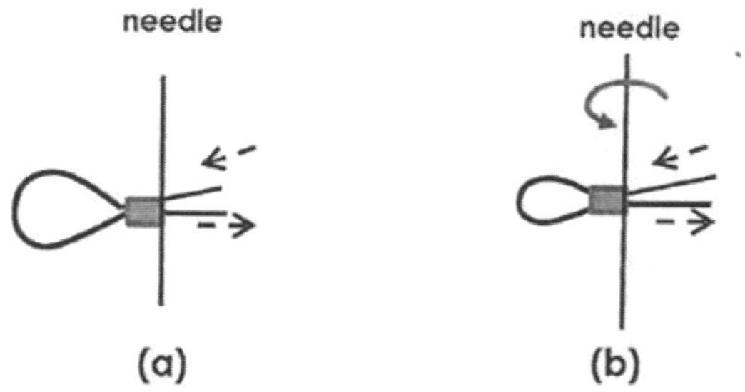

Figure 4. Comparison of bend radius of sensor (a) without the effect of rolling of needle and (b) with the effect of rolling of needle.

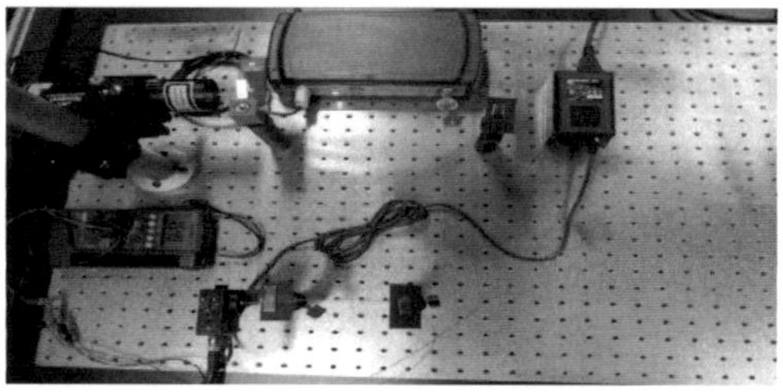

(b)

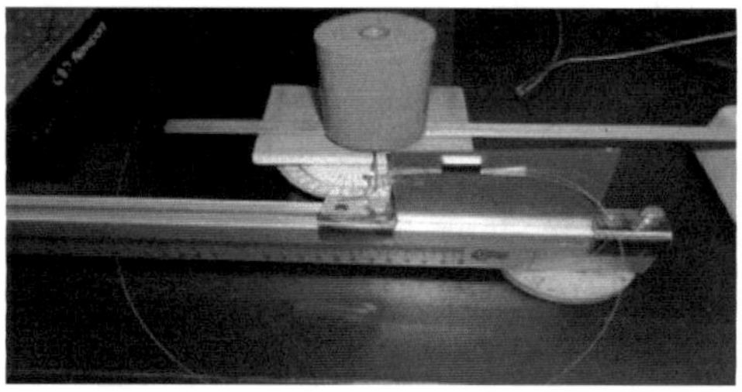

(c)

Figure 5. Photographs of the experimental setup: (a)shown a bar of metal placed on hot plate ,(b) shown exploits of the laser and photodetector and (c) shown installation of the macrobending loss sensor.

IOP Conf. Series: Journal of Physics: Conf. Series **1027** (2018) 012007 doi:10.1088/1742-6596/1027/1/012007

Photograph of the experimental setup was shown in Figure 5. Hot plate provided heat to increase temperature from 30 to 300 degree Celsius. Metal conducted heat and lead to the expansion. When the optical fiber sensor experienced to bending effect due to the expansion of metal, the changes in output light intensity (△I) were recorded. In our demonstration, aluminum, brass and copper bars with the dimensions of 12.80 mm x 3.26 mm x1000 mm were utilized as the sample of metal. The measured coefficient of thermal expansion of those metals would be compared.

In addition, we were very keen to study the influence of the size of the diameter of a needle on the sensitivity of the optical fiber sensor. All experimental results will be displayed in the next section.

3. Results and Discussion

Each figure should have a brief caption describing it and, if necessary, a key to interpret the various lines and symbols on the figure.

3.1 Investigation the coefficient of thermal expansion for the selected metals

In the first experiment, we proposed to compare the thermal expansion of aluminum, brass and copper bars. Here, the needle which composed of stainless steel with fixed diameter of 1.74 millimeter was used for the whole experiment. As the temperature of metal bar increased, the optical loss induced by the bending effect was monitored. For temperature in the range of 150-300 °C, the data points were approximated by a linear curve. For all kind of metal bars, the relationship between the change in optical intensity (△I) and temperature (T) trended to be linear as displayed in Figure 6.

As can be seen, the larger the slope of graph, the more rapid increase of expansion. Here, it should be noted that the slope of graph was described as the coefficient of thermal expansion. According to the measured results and comparing among three metals, aluminum had the highest coefficient of thermal expansion. We obtained values of the coefficient of thermal expansion to be 0.45, 0.35 and 0.32 a.u./°C for aluminum bar, brass bar and copper bar, respectively. It seem to us that brass and copper bars have very similar value of coefficient of thermal expansion. We compared our measured results with the values that published in physics textbook [1]. The value of coefficients of expansion was listed as $25x10^{-6}$, $19x10^{-6}$ and $12x10^{-6}$ $(^{\circ}C)^{-1}$ for aluminum, brass and copper, respectively. Due to the differently measured method, we could just consider arranging the value of the coefficient of thermal expansion of those metals. Thus, our results were in accordance with the results obtained by other method. For now, we were satisfy with the performance of this design system because at least it could resolve the characteristic for each kind of metal.

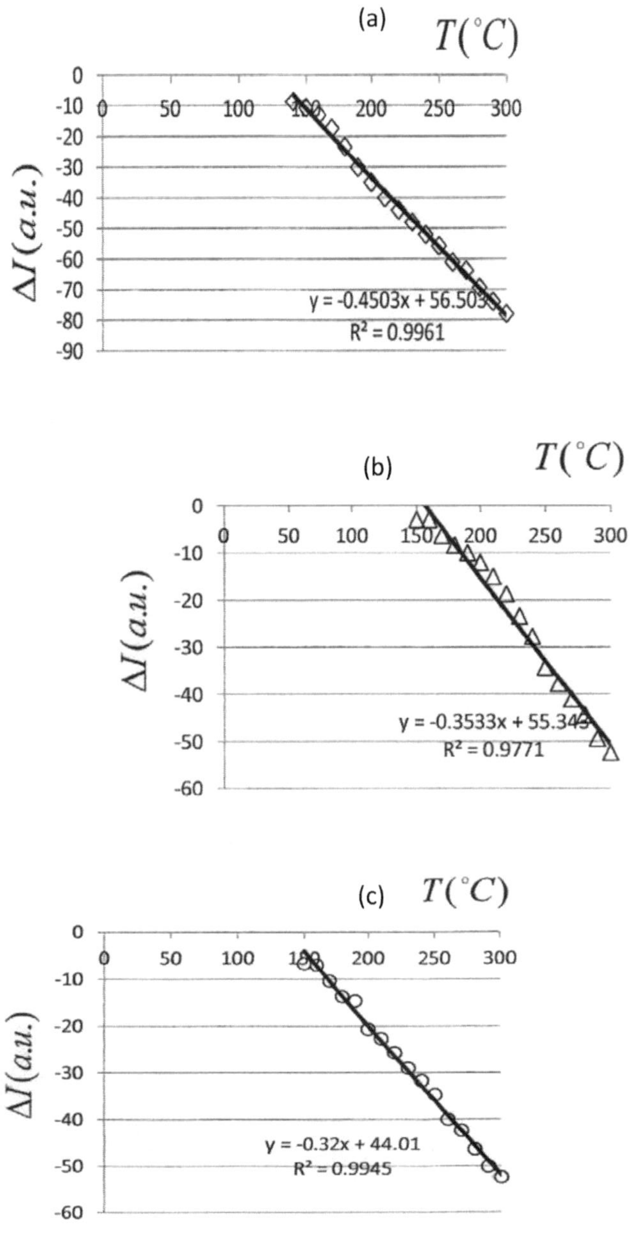

Figure 6. The change in optical intensity (ΔI) as a function of temperature for monitoring thermal expansion of various metals: a) aluminium, (b) brass and (c) copper.

For the future work, it needs to convert the light output reading to the corresponding expanding distance. Also we may further do the calibration between our results with the standard values. We may modify some part of our design sensor to serve as a macrobend displacement sensor which had been presented by Efendioglu et al [9]. Similar work was also adapted by Li et al [10] to monitor corrosion of metal. With that smart system, it is enable us to measure the coefficient of thermal expansion in the unit of length per temperature such as millimeter /°C. In addition, the tapered plastic optical fiber may introduce in the sensing system in order to achieve enhanced performance. Combination of tapering and macrobending ideas are optimized and then it motivates the fiber sensor to reduce the number of modes [11]. Without doubt, it will affect the sensing performance.

3.2 The influence of the size of the diameter of a needle on the sensitivity of sensor

In the second part, only aluminum bar was selected to use and we employed needles with different size. Figure 7 shows the response of sensor to the varying size of the diameter of a needle. Obviously, the sensor shows more sensitive to the larger needle. This means that larger size of needle could contribute to speed up the bending loss of the optical fiber. Attention should be paid to the case of larger size of needle. However, the results which obtained from using needle with diameter of 1.74 mm and 3.90 mm were given quite similar sensitivity. One can predict that once the needle become large enough, the sensitivity of sensor may saturate and limit. For further scope of investigation, a considerable amount of friction between needle and crimp tube must be concerned.

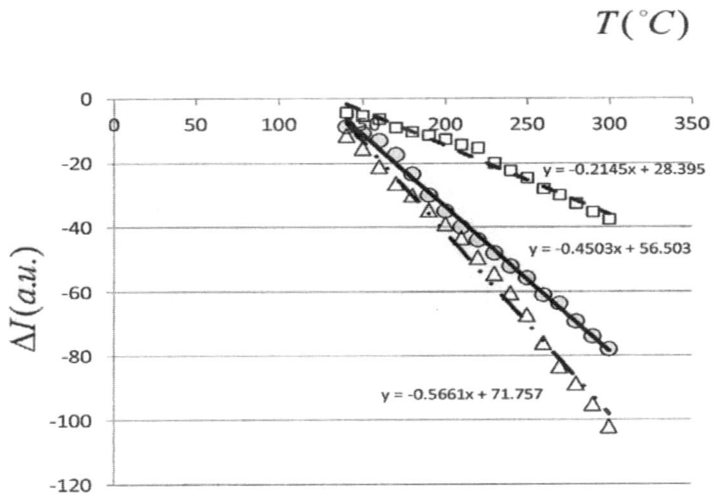

Figure 7. Comparison the change in optical intensity (¯I) as a function of temperature for monitoring thermal expansion of aluminium bar when using various sizes of diameter of needle with the following marks to represent the diameter of needle: □ 1.24 mm, O 1.74 mm and Δ 3.90 mm.

4. Conclusion

An innovation of optical fiber sensor for measuring thermal expansion of metal was proposed. The compact and sensitive sensor used the concept of light loss by elaborating the macrobending loss. The coefficient of thermal expansion of aluminum, brass and copper bars were revealed. Finally, experimental results indicated that the sensitivity of the sensor could be enhanced by increasing the size of the diameter of a needle.

Acknowledgement

This work is supported by the Faculty of Science, Silpakorn University, Thailand.

References

[1] Giancoli, D C 1995 Physics Forth Edition Prentice Hall

[2] Kanagaraj S and Pattanayak S 2003 Measurement of the thermal expansion of metal and FRPs *Cryogenics* **43**(7) 399-424

[3] Wolff E G and Eselun S A 1979 Thermal expansion of a fused quartz tube in a dimensional test facility *Rev Sci Instrum*. **50**(4) 502–506

[4] Grattan K T V and Meggitt B T 2000 Optical fiber technology *Kluwer Academic Publishers*

[5] Pirozzi S 2012 Multi-point force sensor based on crossed optical fibers *Sensors and Actuators* A: Physical **183** 1-10

[6] Liu J, Hou Y, Zhang H, Jia P, Su S, Fang G, Liu W and Xiong J 2017 A wide-range displacement sensor based on plastic fiber macro-bend coupling *Sensors* **17**(1) 196-204

[7] Gattass R, Kung F, Busse L, Shaw L and Sanghera J 2014 Bend loss in multimode chalcogenide fiber at infrared wavelengths *Optical Engineering* **53**(1) 010502-1 –010502-3

[8] Moraleda A, Garcia C, Zaballa J and Arrue J 2013 A temperature sensor based on a polymer optical fiber macro-bend *Sensors* **13**(10) 13076-13089

[9] Efendioglu H S, Sahin A K, Yildirim T and Fidanboylu K (2011, December) Design of hetero-core smart fiber optic macrobend sensors In *Electrical and Electronics Engineering* (ELECO), 2011 7th International Conference on (II-372) IEEE

[10] Li W, Ho S C M, Luo M, Huynh Q, and Song G 2017 Fiber optic macro-bend based sensor for detection of metal loss *Smart Materials and Structures* **26**(4) 1-9

[11] Teng C, Jing N, Yu F and Zheng J 2016 Investigation of a macro-bending tapered plastic optical fiber for refractive index sensing *IEEE Sensors Journal* **16**(20) 1-9 from http://ieeexplore.ieee.org/document/7547294/

IOP Conf. Series: Journal of Physics: Conf. Series **1027** (2018) 012008 doi:10.1088/1742-6596/1027/1/012008

Recycling of pneumatic scrap tyre into nano-crumb rubber by pulsed laser ablation in different pH media

Nur Ezaan Khamsan*[1], Noriah Bidin[1], Shumaila Islam[1], Suzairi Daud[1], Ganesan Krishnan[1], Mohamad Aizat A. Bakar[1], Muhamad Naqiuddin Razali[1], Jamil Khamis[2]

[1]Laser Center and Physics Department of Faculty Science, Universiti Teknologi Malaysia, Skudai, Johor, 81310 Malaysia.
[2]Revocomm Technologies Sdn Bhd., No 11 Jalan Jasmine 1, Bandar Bukit Beruntung, Section BB 10, 48300 Rawang, Selangor Darul Ehsan, Malaysia.

Corresponding author: nur.ezaan@gmail.com

Abstract. Nano crumb rubber from scrap tyre is synthesized via 1064 nm pulsed Nd:YAG laser ablation in three different pH media i.e. DI-water (pH~6.45), D-limonene (pH~3.47) and NaOH solution (pH~13.41). Field Emission Scanning Electron Microscope (FESEM) results show spherical morphology of crumb rubber with high degree of aggregation in DI-water and in D-limonene. However, dispersion of crumb rubbers is observed in NaOH solution. The smallest particles size is obtained in NaOH solution within the range of 10.9 nm – 74.3 nm. Energy-dispersive X-ray spectroscopy (EDX) and FTIR analysis confirmed the elements distribution and chemical bonding of rubber with DI-water, D-limonene and NaOH solution. The experimental findings shows that pulsed Nd:YAG laser ablation has potential for fabricating nano-crumb rubber in liquid media.

1. Introduction

Recently, scrap tyre wasted create adverse effects on the environment because waste tyre disposal process directly contributes towards the enhancement of biodiversity due to its insoluble and toxic compounds [1]. Numbers of researches have been carried out regarding to this particular matter [2-7]. Advent of industrial technology reported that waste tire crumbs have potential as recycled rubber-containing products ranging from construction to road pavement industry [8-10]. To date, there are several processes being practiced for manufacturing of crumb rubber. Two of the most common techniques are ambient grinding [11] and cryogenic processing [12]. However, ambient grinding technique refers to a series of processes that are carried out to yield the desired product and time consuming. Typically, granulation equipment or cracker mills are used in this process. Meanwhile, in cryogenic processing, liquid nitrogen is used to freeze or reduce the temperature of the scrap tyre. By this method, crumb rubber can be produced up to 50 μm in size [12]. However, this technique requires special designed chamber with high intensity instrument to produce finer rubber crumbs.

Crumb rubber manufacturing technologies are continuously focusing on a mechanism to yield finer crumb rubbers from scrap tyre ranging from 40 to 200 mesh. In this study, we proposed a new approach to synthesize finer crumb rubber from scrap tyre by pulsed laser ablation technique in liquid (PLAL). This technique offers green synthesis of nanoparticles or particles' clusters without using any

any toxic chemicals [13]. Several researchers reported the synthesis of nanoparticles from pure metal such as gold [14], zinc oxide [15], organic compound i.e. Eu(TMA) $(H_2O)_4$ [16] using PLAL. Basically, in PLAL technique, nano/or finer particles can be synthesized from hard target (metal) by focusing a pulsed laser on the target surface. High pressure and temperature of plasma as well as a stronger shockwave with a longer duration and a higher impact force leads the formation of finer particles compared to ambient air.

Lately, owing to hybrid approaches, different aqueous liquid solutions e.g., water, ethanol, isopropanol and acetone have been applied to assist laser ablation [17-19]. The D-limonene was chosen because of its environmental friendly nature, non-toxicity and naturally can be found in lemon and orange peels. To the best of our knowledge, we are reporting a new fast, non-toxic and easy approach i.e. PLAL (Nd:YAG 1064 nm laser) within different pH media such as DI-water (pH~6.45), D-limonene (pH~3.47), and NaOH (pH~13.41) to produce the crumb rubber within range of 117 nm – 10 nm.

2. Experimental

Pneumatic scrap tire (which is made of synthetic rubber, natural rubber, fabric, carbon black and C, O, Zn, Si and Si) is used as a target material. The scrap is cut into small piece with dimension of 2 cm x 2 cm x 0.5 cm and then cleaned it in acetone for 30 min using Branson ultrasonic bath and rinsed it thoroughly with de-ionized water. A 1064 nm pulsed Nd:YAG laser is used as a source in laser ablation, operating at a pulse width of 8 ns, repetition rate of 3-Hz, and maximum pulse energy at 311 J. The laser is focused via objective lens at a focusing length of 100 mm onto the rubber surface. The sample is immersed in 10 ml of DI-water (pH~6.45), D-limonene (pH~3.47) and NaOH solution (pH~13.41). After this, each solution is rotated individually by rotational plate (Newport®-UTR80) at 120 rpm driven by a stepping motor to avoid crater formation during the ablation processes. The solutions are processed within different time intervals i.e. 10 min, 20 min, 30 min, 40 min and 50 min at room temperature. Subsequently, variations in solutions color are observed by increasing/changing the pulse numbers. Figure 1 shows the schematic diagram of pulsed laser ablation in liquid to produce the crumbs rubber.

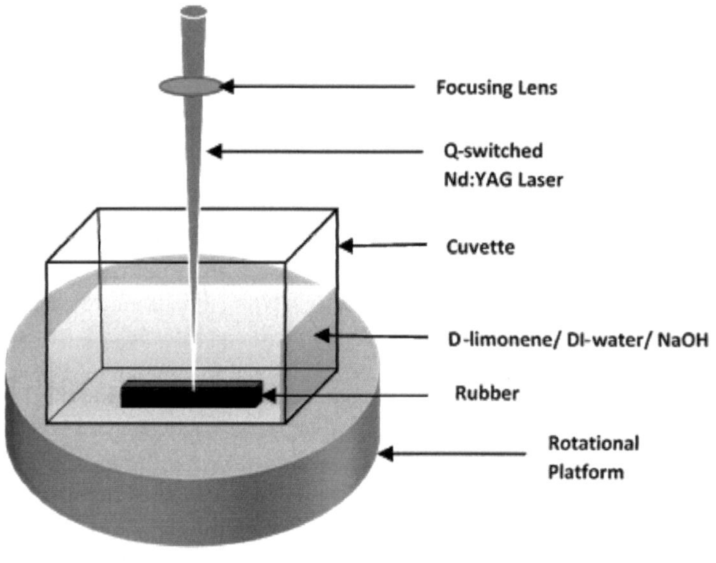

Figure 1. Schematic diagram of pulsed laser ablation for crumbs rubber producing from scrap tyre.

3. Characterization

Morphology of synthesized crumb rubber was observed using Field Emission Scanning Electron Microscope (FESEM) HITACHI SU8020. For FESEM analysis, one drop of each colloidal solutions was placed on 15 mm x 15 mm x 2 mm FTO glass slide. Each of the glass slide was then baked in an oven at 200 °C for 8 min. Then, the glass slides were coated using platinum by automated platinum sputter (QUOROM Q150RS) for about 5 min. The samples were examined through FESEM, operating at ultra-low accelerated voltage imaging resolution of 1.3 nm at 2.0 kV. The elemental analysis was also carried out by an energy-dispersive X-ray spectroscopy (EDX) spectra (Oxford Silicon Drift Detector
[SDD]). The infrared spectra were recorded using a frontier FTIR (ATR) spectrometer, Perkin Elmer Series L160000A (USA) within the range of 600 to 4000 cm^{-1} at ambient temperature.

4. Results and Discussions

Digital pictures in Figure 2 (A, B, C) shows that the color varies in different solutions i.e. DI-water, D-limonene and NaOH within different ablation time intervals 10 min – 50 min. In DI-water [Figure 2-A (a – e)], concentration of crumb rubber particles is low even after increasing the ablation time as compared to D-limonene and NaOH. Whereas, [Figure 2-B (a - e)] shows that the concentration of crumb rubber particles increases by increasing the ablation time and gathered at the bottom of the falcon tube. After 50 min, high concentration of crumbs was obtained which can be clearly seen in [Figure 2-B (e)]. However, the crumb rubber particles are found to be dispersed in NaOH solution and the concentration of particles increases gradually after increasing the ablation time 10 min – 50 min [Figure 2-C (a – e)].

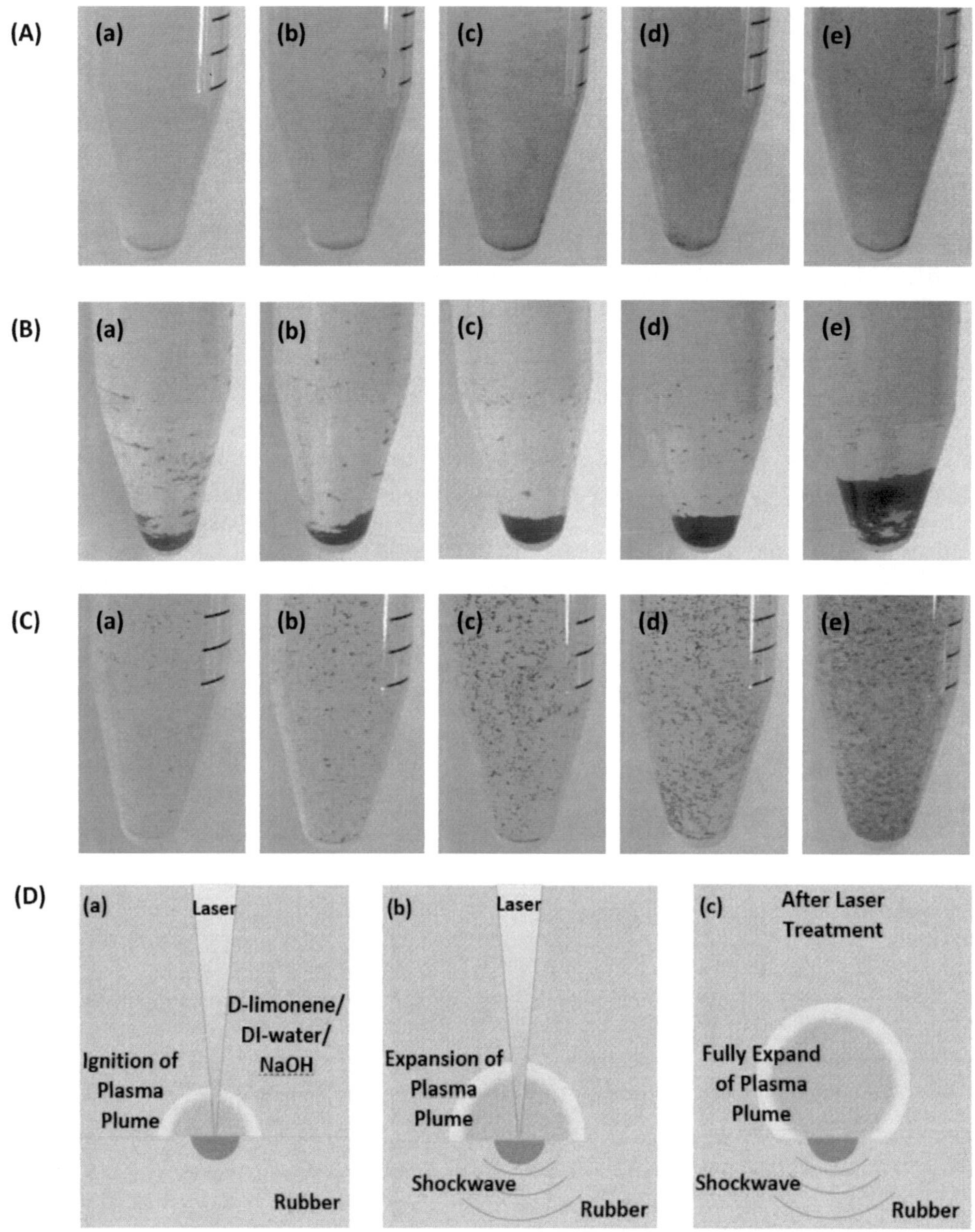

Figure 2. Crumb rubber into (A) DI-water (B) D-limonene (C) NaOH solutions, a, b, c, d and e is corresponding to 5 different ablation times 10 min, 20 min, 30 min, 40 min and 50 min, respectively. Whereas, (D) is the schematic illustration of laser processing, (a) plasma plume is ignited on the irradiated spot, (b) isothermally expansion of plasma plume by laser energy absorption, (c) plasma plume adiabatically expands and chemical reaction takes place at the liquid-material interface after laser pulse treatment.

From figure 2-B (a - e), it is observed that the production of crumb rubber is higher in D-limonene medium as compared to DI-water and NaOH solutions. The enhancement of pulsed laser ablation in d-limonene solution is due to the strong confinement of plasma plume [Figure 2-D (a-c)] and the assisted localized chemical etching forces that occurred within the surrounding medium [20,21]. In mechanism of PLAL technique, focusing of high intensity laser beam on the sample surface (rubber) leads to excitement of electrons to a higher state as a result from inverse-bremsstrahlung process [22]. This process further yields a dense cloud of atoms and ions in excited state (C^+, H^+, O^{2-}) due to energy transfer from electron-ion and electron-neutral collision processes. Accordingly, transfer of energy from the laser pulse formed the plasma plume [Figure 2-D (a)] that consists a mixture of vapours containing those ions, atoms and electrons all together. The plasma plume then experienced an isothermal expansion due to continuous absorption from laser pulse energy [Figure 2-D (b)], resulting in relatively high speed increment in temperature and pressure. After the laser treatment, plasma plume continues to expand adiabatically with relatively high pressure and temperature [Figure 2-D (c)]. This expansion last for about few microseconds before it shrinks up and collapse. Under this extreme condition, thermal evaporation and decomposition of molecules, are excited, resulting in the generation of cavitation bubbles. Energetic expansion of plasma plume induced a shock wave with a relatively strong pressure. The plasma plume condensates into clusters and aggregates into smaller particles [23]. The smaller particles (ejected materials) distributed in the irradiated area. Furthermore, the chemical reaction around the irradiated area further increases the removal of target rubber. The size and the weight of the removal particles are apparently greater than the rest of other tested media, consequently the precipitation of the crumb rubber is obviously observed from the D-limonene solution.

Figure 3 shows the graph of crumb rubber concentration in three different solutions i.e. DI-water, D-limonene and NaOH versus ablation time. It can be observed that the concentration of rubber particles increases by the ablation time (10 min – 50 min). The D-limonene solution curve (b) has the highest concentration followed by NaOH solution curve (c) and the lowest concentration is in DI-water curve (a). The concentration of crumb rubber in D-limonene solution is higher than the DI-water and NaOH is due to the precipitation of rubber particles at the bottom of the container which also confirmed as shown in Figure 2(B). The lowest concentration is observed in DI-water because less crumb rubber particles produced in the solution. The average concentration of crumb rubber in D-limonene is 0.2211 g/mL, NaOH with 0.0346 g/mL, meanwhile for DI-water 0.0011g/mL.

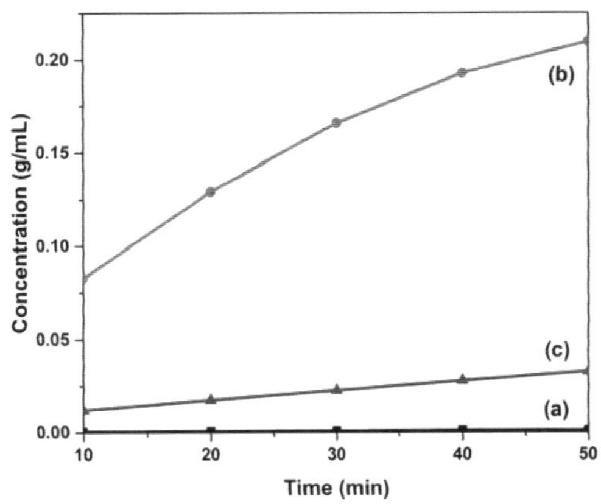

Figure 3. Crumb rubber concentration in different solutions (a) DI-water, (b) D-limonene and (c) NaOH solution, versus ablation time.

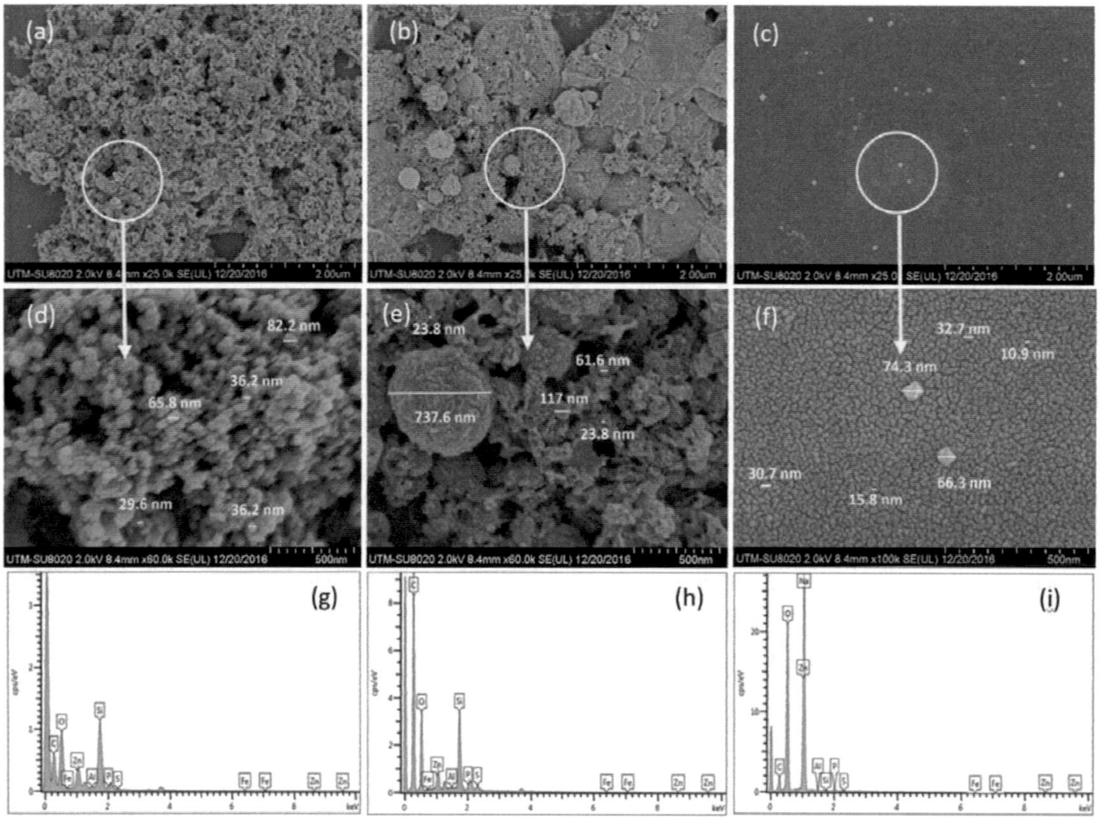

Figure 4. FESEM images of crumb rubber in (a) DI-water, (b) D-limonene, and (c) NaOH solution, (d, e, f) magnified images, whereas, (g, h, i) EDX spectrums corresponding to (a, b, c), respectively.

Table 2. EDX analysis of crumb rubber in different media

Crumb rubber Media	Elemental composition (W %)							
	C	O	Si	Zn	S	Fe	Al	Na
De-ionized water	43.8	24.8	19.9	9.7	0.9	0.5	0.4	-
D-limonene	65.3	21.8	8.6	2.8	0.8	0.5	0.2	-
NaOH	11.7	41.2	0.3	8.1	0.5	0.3	0.1	37.8

The morphology and particles size of laser ablated crumb rubber after 50 min treatment were analysed by FESEM. Figure 4 (a, b, c) shows morphology of rubber particles in DI-water, D-limonene and NaOH solution, respectively. Figure 4 (a) shows spherical shape particles' aggregation in DI-water is probably due to absorb water molecules on the surface of the crumb rubber. Meanwhile, Figure 4 (d) is the magnified image of Figure 4 (a). The observed particles size is in the range from 29.6 – 82.2 nm. Moreover, cluster formation (737.6 nm) with particles aggregation can be clearly observed in D-limonene as shown in Figure 4 (b) and its magnification for the circular area is shown in Figure 4 (e). Particles size are observed in the range of 23.8 – 117 nm. Particles size within the range of 10.9 – 74.3 nm are realized in NaOH solution as depicted in Figure 4 (c), and it corresponding magnification is shown in Figure 4 (f). The hydroxide ions from NaOH solution chemically break into mono-, di- and polysulfidic cross-linkages in rubber vulcanizate, resulted in small particles dispersion of crumb rubber [24].

Figure 4 (g), (h) and (i) shows the EDX spectrum of laser ablated crumb rubber in DI-water, D-limonene and NaOH solutions. The element composition in weight percentage (wt. %) of crumb rubber from each solution is tabulated in Table 2. The main component of crumb rubber is comprised of carbon C, oxygen O, silicon Si, zinc Zn, iron Fe, sulphur S and aluminium Al. The EDX analysis, suggest that there are no foreign elements detected from fabricated crumb rubber. However, the presence of sodium Na in Figure 4(i) is attributed from the NaOH solution. This finding indicates that pulse laser ablation in liquid technique has ability to induce crumb rubber.

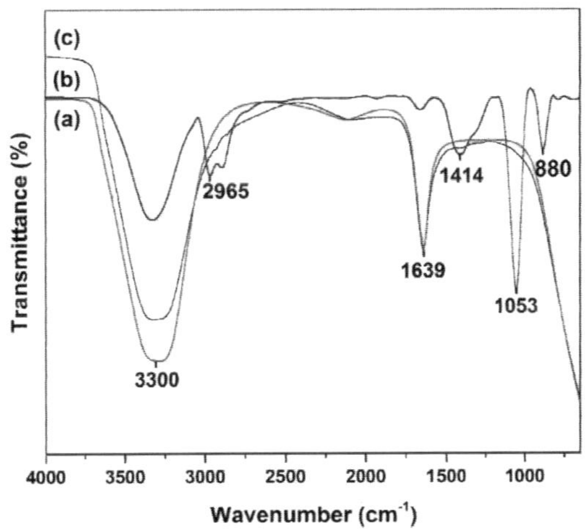

Figure 5. FTIR spectra of laser ablated crumb rubber in (a) DI-water, (b) D-limonene and (c) NaOH

The broad absorption band at ~3300 cm^{-1} associated with hydroxyl group stretching in all three samples. A little intense peak around ~2965 cm^{-1} [Figure 5(a)], is due to C-H symmetric stretching of organic residue. Occurrence of sharp absorption band in all three samples around ~1639 cm^{-1} is characteristic of a C=O group [25]. The prominent peak at ~1053 cm^{-1} is attributed to the bending of =C-H bonds that indicates the presence of D-limonene in the solution [Figure 5(b)]. Prominent bands at ~1414 cm^{-1} and ~880 cm^{-1} correspond to the CH_3 and =CH out of plane bending of cis-1,4-polyisoprene the component in natural rubber [26]. FTIR analysis confirmed that the rubber species have good interaction with different pH media.

5. Conclusion

Nano-crumb rubber (10 – 117 nm) was synthesized using PLAL technique in different pH media. The concentration of crumb rubber is found to be optimum in acidic solution like D-limonene with pH~3.47. The smallest nanoparticle of rubber is achieved when the crumb rubber is synthesized using basic solution like sodium hydroxide NaOH with pH~13.41. Beside the crumb rubber particles are realized to be more homogenized in this basic solution. However, the crumb rubber are very difficult to be achieved using deionized water (pH~6.45). FESEM images show the spherical morphology of crumb rubber within the range of particle size (DI-water, 29.6 – 82.2 nm, D-limonene, 23.8 – 117 nm, NaOH solution, 10.9 – 74.3 nm). EDX analysis confirms the presence of crumb rubber in the solutions from scrap tyre. FTIR analysis shows the chemical bonding between three different pH media (DI-water, D-limonene, and NaOH) and rubber species. The experimental findings demonstrated that

PLAL is a promising technique to fabricate nano-crumb rubber from a recycle scrap tyre compared to mechanical grinding methods.

Acknowledgement
We would like to thank the Government of Malaysia for the financial support through GUP Tier 1 15H52 and Universiti Teknologi Malaysia. Thanks, are also due to Mybrain15 MyPhD for the scholarship sponsor for the first author.

References

[1] B S Thomas and R C Gupta 2016 A comprehensive review on the applications of waste tire rubber in cement concrete. *Renew. Sustainable Energy Rev.* **54** 1323-1333

[2] O Holst, B Stenberg and M Christiansson 1998 Biotechnological possibilities for waste tyre-rubber treatment *Biodegradation* **9** 301-310

[3] A M Mastral, R Murillo R, M S Callen and T Garcia 2000 Optimisation of scrap automotive tyres recycling into valuable liquid fuels *Resources, Conservation and Recycling* **29**(4) 263– 272

[4] B Wu and M H Zhou 2009 Recycling of waste tyre rubber into oil absorbent *Waste Management* **29**(1) 355–359

[5] R Murillo, E Aylo´n, M V Navarro, M S Calle´n, A Aranda and A M Mastral 2006 The application of thermal processes to valorise waste tyre *Fuel Processing Technology* **87**(2) 143–147

[6] Chitsan Lina, Chun-Lan Huang and Chien-Chuan Shern 2008 Recycling waste tire powder for the recovery of oil spills *Resources, Conservation and Recycling* **52** 1162–1166

[7] M C Bignozzi and F Sandrolini 2006 Tyre rubber waste recycling in self-compacting concrete *Cement and Concrete Research* **36**(4) 735–739

[8] Fabiana Maria da Silva, Luisa Andréia Gachet Barbosa, Rosa Cristina Cecche Lintz and Ana Elisabete P.G.A. Jacintho 2015 Investigation on the properties of concrete tactile paving blocks made with recycled tire rubber *Construction and Building Materials* **91** 71-79

[9] Adam J. Kardos and Stephan A. Durham 2015 Strength, durability, and environmental properties of concrete utilizing recycled tire particles for pavement applications *Construction and Building Materials* **98** 832-845

[10] Matteo Pettinari and Andrea Simone 2015 Effect of crumb rubber gradation on a rubberized cold recycled mixture for road pavements *Materials & Design* **85** 598-606

[11] Avraam I. Isayev 2013 Recycling of Rubbers In: The Science and Technology of Rubber (Fourth Edition) *Academic Press Boston* 697-764

[12] S B Liang and Y C Hao 2000 A novel cryogenic grinding system for recycling scrap tire peels *Advanced Powder Technology* **11**(2) 187-197

[13] Takeshi Sasaki, Yoshiki Shimizu and Naoto Koshizaki 2006 Preparation of metal oxide-based nanomaterials using nanosecond pulsed laser ablation in liquids *Journal of Photochemistry and Photobiology A: Chemistry* **182**(3) 335-341

[14] Wan Norsyuhada Wan Shukri, Noriah Bidin, Shumaila Islam, Ganesan Krishnan, Mohamad Aizat Abu Bakar and Mohd Syafiq Affandi 2016 Synthesis and characterization of uncoated and cysteamine-coated gold nanoparticles by pulsed laser ablation *J. Nanophoton* **10**(4)

[15] Khaldoon N. Abbas, Noriah Bidin and Raad S. Sabry 2016 Controllable ZnO nanostructures evolution via synergistic pulsed laser ablation and hydrothermal methods *Ceramics International* **42**(12) 13535–13546

[16] Ohanna Maria Menezes Madeiro da Costa and W.M. de Azevedo 2016 Highly luminescent metal organic framework Eu(TMA) (H2O)4 materials prepared by laser ablation technique in liquid *Journal of Luminescence* **170**(2) 648-653

[17] M E Shaheen, J E Gagnon and B J Fryer 2013 Femtosecond laser ablation of brass in air and liquid media *J. Appl. Phys.* **113**(21)

[18] N Krstulović, S Shannon, R Stefanuik, and C Fanara 2013 Underwater-laser drilling of aluminium *Int. J. Adv. Manuf. Technol.* **69** 1765–1773

[19] S Zhu, Y F Lu and M. H. Hong 2001 Laser ablation of solid substrates in a water-confined environment *Appl. Phys. Lett.* **79**(9) 1396–1398

[20] Shen MY, C H Crouch, J E Carey and E Mazurc 2004 Femtosecond laser-induced formation of submicrometer spikes on silicon in water *Applied Physics Letters* **85**(23) 5694-5696

[21] Kumar B and R K Thareja 2012 Growth of titanium nanoparticles in confined plasma *Physics of Plasmas* **19**(3) 033516

[22] Aghaei M, S Mehrabian and S H Tavassoli 2008 Simulation of nanosecond pulsed laser ablation of copper samples: A focus on laser induced plasma radiation *Journal of Applied Physics* **104**(5) 053303

[23] Haibo Zeng, Weiping Cai, Yue Li, Jinlian Hu and Peisheng Liu 2005 Composition/Structural Evolution and Optical Properties of ZnO/Zn Nanoparticles by Laser Ablation in Liquid Media *The Journal of Physical Chemistry B.* **09**(39) 18260-18266

[24] Myers R D, Nicholson P, Macleod J B and Moir M E 1997 Rubber devulcanization process: *Google Patents*

The electrical conductivity and energy band gap of '*bunga belimbing buluh*'/tio$_2$ nanocrystals as hybrid solar cell

N H Kamarulzaman[1], H Salleh[*2], M S M Ghazali[3], S M Ghazali[4] and Z Ahmad[5]

[1,3]School of Ocean Engineering, [2]Centre for Fundamental and Liberal Education, [3,5]Advance Nano-materials Research Group, School of Fundamental Science, Universiti Malaysia Terengganu, 21030 Kuala Nerus, Terengganu, Malaysia

Corresponding author: hasiah@umt.edu.my

Abstract. This research intends to explore the effect of thickness of inorganic titania nanocrystals (TiO_2 NC_s) materials and *Averrhoe bilimbi*'s flower towards the electrical conductivity. *Averrhoe bilimbi's* flower or also known as 'bunga belimbing buluh' was used for the first time as a natural dye in hybrid solar cells. The performance of electrical conductivity can be improved in bilayer heterojunction hybrid solar cell (HCS). The TiO_2 NC_s was deposited on the ITO substrate using Electrochemistry method at room temperature. The dye extracted from *Averrhoe bilimbi's* flower was deposited on the top of TiO_2 NC_s layered using the same method. The electrical conductivity can be recorded using Four Point Probe (FPP) under dark and light radiation (range of 0 Wm^{-2} to 200Wm^{-2}). From the results, electrical conductivity was increased by the increment light intensity and suitable for further solar cell fabrications.

1. Introduction

Renewable energy is any source that generated from natural resources such as wind, solar geothermal and hydropower. The search for clean and renewable energy sources has become one of the greatest challenges in our society, due to the rapid depletion of fossil fuels and increasing demand on energy supply. One of the most promising alternative energy sources is solar energy which is in exhaust, purity, cheap and flexible use [1]. Solar cell is one of the potential candidates for solving the present world's energy crisis. The solar cells used in harvesting the solar power are commonly categorized into different types with respect to the composition of their material such as organic dye solar cells, hybrid solar cell, non-crystal, multiple crystal, and single crystal silicon solar cells. In a few years, hybrid solar cell (HSC) has drawn attention as an alternative to traditional silicon or other inorganic technologies. HSC is the combination of both organic and inorganic semiconductors [2].

The advantage of organic material are good process ability with low cost deposition techniques, environmental, cost-effective, mechanical properties of the constituent materials, including the use of plastic-based substrates and encapsulates, open the possibility of manufacturing flexibility and low weight solar modules [3]. Foremost, the natural dyes abundantly available in plant parts, like flowers, seeds, barks, leaves, and stem can be extracted by simple procedure. Due to their cost efficiency, easily extractable using cheap organic solvents, non-toxicity, and environmental friendly, natural dyes are still a popular research [4].

Solar energy with natural dyes is not fully utilized in Malaysia. This research will be conducted to reduce energy consumption and the data and knowledge acquired will assist in the development of hybrid solar cell specifically for Malaysian climatic conditions. In this study, *Averrhoe bilimbi*'s flower was used for the first time as a natural dye in hybrid solar cells. Therefore, this research attempt to express the significant impact of Averrhoe bilimbi's flower with Titanium dioxide photo-anode in bilayer heterojunction. Varied thicknesses of TiO_2 coating occurred with different techniques in Electrochemistry Impedence Spectroscopy (EIS). TiO_2 semiconductor coating thickness has favourable impact to the photovoltaic response to adsorbs more dye molecules [5, 6]. Futhermore, this combination absorbed a wide range in solar spectrum.

In Malaysia, the fruit of Belimbing buluh or known as *Averrhoe bilimbi* is very popular and commonly used in Malay cuisine. Every part of this plant has its own benefits. Therefore, the dye used in this research was extracted from the flower of this plant. This *Averrhoe bilimbi* prefers direct sunlight and seasonally humid climate [7]. The flower begins to bloom in yellowish green or purplish marked with dark-purple colour around February and continuously blooms and fruitful until December [7]. The flower was chosen because it is widely used in traditional medicine to treat many symptoms and it is said to be effective against coughs and thrush [7-9]. Figure 1 and 2 shows the *Averrhoe bilimbi's* flower and the structure of *Averrhoe bilimbi.*

Figure 1. Averrhoe bilimbi's flower

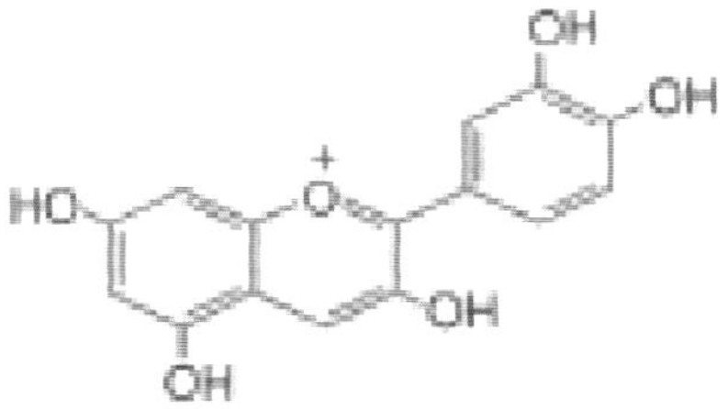

Figure 2. Chemical structure of cyanidin [10]

Anthocyanin generally found in most species of plant such as in fruits, flower and vegetables. Anthocyanins are one of the derived structures taken from flavanol. Examples of anthocyanidins are delphinidin (blue to red), pelargonidin (orange) and cyanidin (orange to red) [9]. The colours and stability are affected by substitution patterns and the glycocyl group of anthocyanin. The natural pigment compounds in many fruits and flowers can be determined by the colours. Red, blue and purple are the colours usually appear in anthocyanins [11]. Anthocyanins are most distinguished by their colour range,which is determined by the wavelength. The wavelength for Averrhoe bilimbi's flower in visible range fall between 400 nm and 650 nm and also absorb under acidic condition.

Intensity and type of colour was affected by hydroxyl and methoxyl group [10]. Methoxyl group was determined by red colour and hydroxyl group go to blue colour.

2. Experimental

The organic dye was extracted from nature plant and this research was conducted in few stages.

2.1 Sample Collection and Extraction of Belimbing buluh's flower

Averrhoe bilimbi's flower was collected from a small village in Dungun, Terengganu. The fruits are red in colour. After that, the flower was transported to the laboratory on the same day. It was cleaned using tap water and then rinsed with distilled water. Then, it was crushed into fine powder before pouring into 100 ml ethanol. The mixture was kept for one week at room temperature. Subsequently, the solution was filtered using filter paper and dark red extract was collected. Then, the solution was placed in ultrasonic bath for 10 minutes at temperature 30°C. The extraction was concentrated and used as a dye to TiO_2 nanoparticles for the fabrication of solar cell [12., 13]. Ultrasonic bath was used to maximize the efficiency of natural colorant extraction [14]· The obtained samples were fabricated to be used as solar cell and tested in term of electrical efficiency.

2.2 Preparation of Indium Tin Oxide (ITO) Substrate

Indium tin oxide (ITO) coated glass was used as a substrate. The ITO glass was cleaned using ultrasonic bath (JEIOTECH, United Kingdom) to avoid dirt and dust. This step is very important to reduce any impurities present in the ITO and also to avoid contamination [2]. The ITO coated glass substrates which used as working electrode was cleaned with detergent solution for 10 minutes in ultrasonic bath at 30°C. Then, the ITO coated glass was put in distilled water for 10 minutes. Lastly, ITO was cleaned with acetone [2] for 10 minutes before final cleansing with the distilled water for another 10 minutes. After that, ITO coated glass substrate was dried using the hair dryer before it was kept in a Petri dish.

2.3 Fabrication of Hybrid Solar Cell

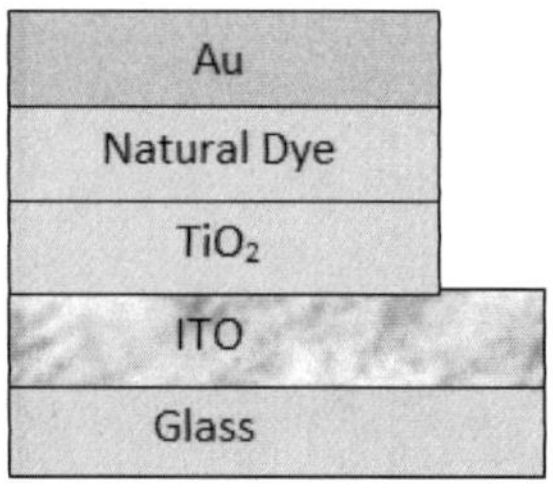

Figure 3. Structure of bilayer heterojunction hybrid solar cell

Based on Figure 3, the thin films were prepared using the electrochemistry method. The first and second layer was prepared using the same method. Electrochemistry method was used for depositing of TiO_2 thin film. Due to the deposition process, the TiO_2 thin films were subjected to a sonication treatment. The suitable temperature was recorded at 450°C for 120 minutes. The treatment will improve the electronic contact between TiO_2 nanoparticles. Then, it was rinsed with distilled water and ethanol. The second layer consisted of organic dye that was deposited on the thin film layer. Lastly, a thin layer of gold (Au) was deposited on the natural dye and it was used as counter electrode.

2.4 Electrical properties characterization

For this research, four point probes (FPP Jandel RM3 Test Unit) were used to determine the electrical conductivity of hybrid solar cell according to equation (1). The two outer probes supply a voltage difference that drives a current through the film. The two inner probes picked up a voltage difference and the sheet resistivity is calculated from Equation (2) via a physical model of the current distribution. R_S is the sheet resistance, 4.532 is the correction factor, V is the voltage measured and I is the current applied from the test unit. The unit of sheet resistance is ohms per square (Ω/m^2):

$$\sigma = I/R_S \quad (1)$$

where the $R_S = 4.532 \text{ x } V/I$

Ultraviolet-Visible (UV-Vis) Spectrometer Model Perkin Elmer Lambda 25 was used to measure the absorption of the spectrum light. Then it passed through a sample and the measurement was recorded after reflection from a sample surface. The selected wavelength range was analyzed between 200-1000 nm in the ultraviolet (UV) and visible (Vis) regions of the electromagnetic spectrum. The data was recorded from Absorption (A) versus wavelength (λ). The dye sensitised solar cells were illuminated using a 75 W Halogen lamp with an incident power of about 100 mW/cm^2 in an illumination area of 0.5 cm^2; UV and IR filter glasses were used in front of the sample [15]. Fourier transform Infra-red (FTIR) spectroscopy model IRTracer-100 was used to determine certain functional groups of pure compound and to confirm the presence of specific impurities. Fourier Transform Infrared spectra of the extracts dyes were recorded in the range of 650-4000 cm^{-1} [16].

2.5 Thickness measurement
The thickness measurement was measured for different natural dyes using Dektak 150 Stylus Surface Profiler at Research center, Universiti Kebangsaan Malaysia (UKM).

3. Results and Discussion

3.1 Absorption Spectra of Natural Dyes (UV-Vis Analysis)
Figure 4 below shows the absorption spectrum of TiO_2 and it was recorded in the range of 250 nm to 320 nm. Wu et al stated that wavelength of TiO_2 was recorded 380 nm smaller in the ultraviolet (Uv) light spectrum [17]. In this research, the absorption spectrum range for *Averrhoe bilimbi*'s flower was recorded in the range of 320nm to 600 nm. *Averrhoe bilimbi*'s flower has a broad and strong absorption in the uv region at 320 nm and blue region at 438 nm with the maximum peak recorded at 350nm. Figure 5 shows the UV-Vis absorption of natural dye solution extracted from *Averrhoe bilimbi*'s flower.

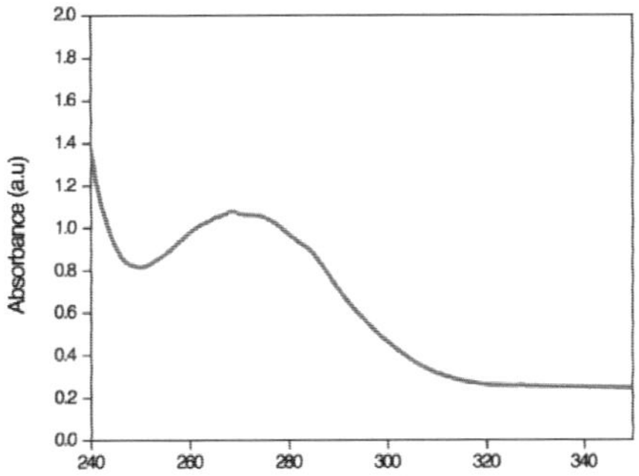

Figure 4. UV-Vis absorption of TiO_2

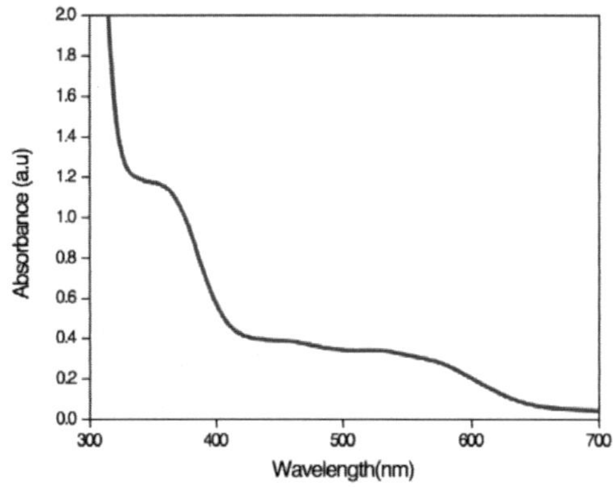

Figure 5. UV-Vis absorption of natural dye solution extracted from *Averrhoe bilimbi*

3.2 Dye Structure (FTIR Analysis)

Figure 6 shows the FTIR spectra of *Averrhoe bilimbi*'s flower. The peak was assigned at 1440 cm^{-1} and 1319 cm^{-1} corresponded to C-H bond ortho-aromatic ring. The absorption peaks at 1645 cm^{-1} was recorded to the functional groups of stretching carbonyl C=C. Meanwhile, the peak at 2945cm^{-1} and 3335cm^{-1} were correlated with the –OH stretching modes.

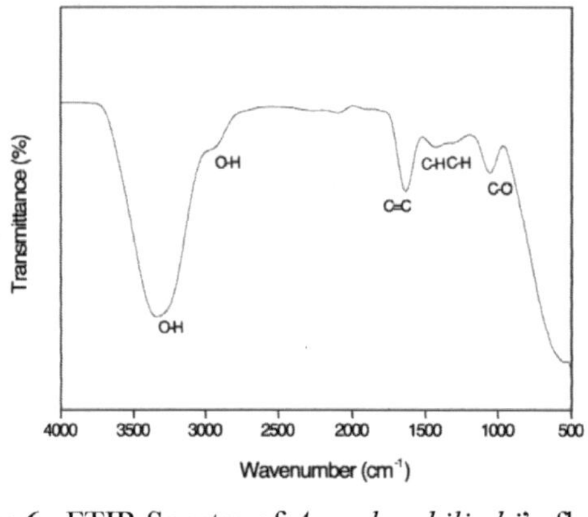

Figure 6. FTIR Spectra of *Averrhoe bilimbi*'s flower

3.3 Thickness of bilayer heterojunction

The absorption spectra of thin films were recorded at different number of scans as shown in Figure 7. The number of scan or cyclic was determined in the different thickness of layers. It can also be seen in the Table 1. The thickness of TiO_2 films were measured by using Profilometer (Bruker, Detakk Profilometer). The number of scan for the thin films was fixed at one, three, five and seven scan. From Figure 7, ITO/TiO_2(5) has the highest UV-Vis absorption compared to others. As can be seen, it was recorded in the range of 250 nm to 320 nm. The thickness of TiO_2 films were recorded in Table 1.

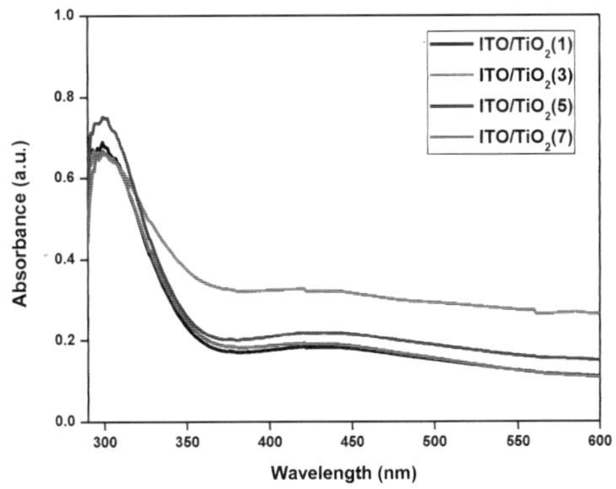

Figure 7. UV-Vis absorption of TiO_2 thin film sample

Table 1. Thickness of the bilayer thin film sample

Number of Scan	Thickness (± 0.1 nm)
1	59.90
3	61.60
5	84.09
7	106.34

3.4 Electrical Conductivity

The electrical conductivity can be determined using Four Point Probe (FPP) under dark and light radiation. From the results, electrical conductivity was increased by the increment light intensity. Figure 8 shows the conductivity of ITO/TiO_2/AB at different number of scans.

Analysis shows the electrical conductivity was recorded between 0.032 Wm^{-2} to 0.045 Wm^{-2}. The highest conductivity was obtained at 0.045 Wm^{-2} for sample ITO/TiO_2 (7)/AB. The lowest conductivity recorded at ITO/TiO_2(3)/AB. The electrical conductivity was rapidly increased from dark condition at 0 Wm^{-2} to light condition which is 200 Wm^{-2}. This is because the samples absorbed light energy and convert it into electrical energy. For dark condition, ITO/TiO_2(3)/AB shows the lowest electrical conductivity while for the highest conductivity measured for ITO/TiO_2(7)/AB under 200 $Wm^{-2}$of light illumination. The increment of intensity of light affected the electrical conductivity. The colorant was affected by adding the layers. The polaron to be localized by 93% at one W-atom [18,19].

When the TiO_2NCs was irradiated, the conductivity changed at different intensity. This was due to the thermal activated mechanism added to photon mechanism [20]. The performance of electrical conductivity can be improved in bilayer heterojunction. This can be explain by the TiO2 semiconductor coating thickness. Futhermore, this combination absorbed a wide range in solar spectrum. As the thickness of TiO_2 increased, the electrical conductivity of the thin film also was increased.

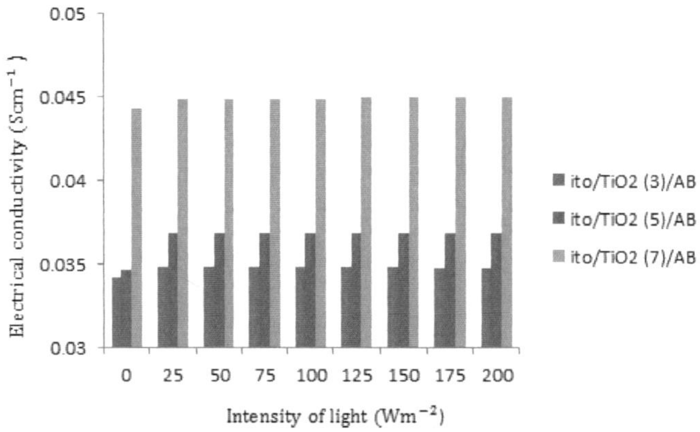

Figure 8. The electrical conductivity of ITO/TiO_2 /AB at different thickness

4. Conclusion

In summary, the electrical conductivity was increased by the increment light intensity. The increment of layer or thickness into surface will improve the absorbance of the sample. The ITO/ TiO_2/ dye as Bilayer Heterojunction Hybrid Solar Cell deposited on the different thickness was successfully prepared by electrochemistry method. The effect of thickness on electrical conductivity bilayer thin film was studied. The highest conductivity was obtained at 0.045 Wm^{-2} for sample ITO/TiO_2(7)/AB. The lowest conductivity recorded at ITO/TiO_2(3)/AB. The ITO/TiO_2(7)/AB thin film shows higher electrical properties. The presence of carbonyl and hydroxyl groups in the anthocyanin molecules bound to the surface of ITO is in favor to facilitate the photoelectric conversion effect. From results obtained, we conclude that our samples deposited on ITO substrate are successfully investigated and improve the electrical properties of the commercial ITO substrate. Furthermore, the study of these dyes is promising and can promote additional studies oriented to educate people on renewable energy sources, and disseminate knowledge for the optimization of solar cell components compatible with such dyes.

Acknowledgements

The authors would like to thank Universiti Malaysia Terengganu (UMT), UiTM for the scholarship SLAB Ministry of Education Malaysia, Fundamental Physics Laboratory UMT's members and Nano Research Team (UMT) for supporting for this research project.

References

[1] Zhang J C, Han Z Y, Li Q Y, Yang X Y, Yu Y and Cao W L 2011 N, S-Doped TiO2 Anode Effect on Performance of Dye-Sensitized Solar Cells. *Journal of Physics and Chemistry of Solids* **72**(11) 1239-1244

[2] Aziz N N, Isa, M.I.N and Hasiah S 2014 Electrical and Hall Effect Study of Hybrid Solar Cell. *Journal of Clean Energy Technologies* **2**(4)

[3] Narayan M R 2012 Review: Dye Sensitized Solar Cells Based On Natural Photosensitizers. *Renewable and Sustainable Energy Reviews* **16**(1) 208-215

[4] Laily A R N, Hasiah S, Aziz N N, Nazri A and Sabri M M G 2016 Significance of Substrate Temperature on Electrical Conductivity, Hall Effect, and Thickness of Bilayer Heterojuction Organic Solar Cell using Rhodomyrtus Tomentosa (aiton) Hassk and Ixora Coccinea l Dye. *Jurnal Teknologi* **78**(3) 67-74

[5] Ruhane T A, Islam M T, Rahaman M S, Bhuiyan M M H, Islam J M M, Bhuiyan T I, Khan K. A and Khan M A 2017 Impact of Photo Electrode Thickness and Annealing Temperature on Natural Dye Sensitized Solar Cell. *Sustainable Energy Technologies and Assessments*

20 72-77

[6] Arman S and Hoda Nourmohammadi M 2015 The Effect of Drying and Thickness of TiO2 Electrodes on the Photovoltaic Performance of Dye-Sensitized Solar Cells. *International Journal of Electrochemical Science* **10** 3354-3362

[7] Wan Norhana M N, Azman M N, Susan E P, Hilton C D and Dykes G A 2009 Effects of Bilimbi (Averrhoa bilimbi L.) and Tamarind (Tamarindus indica L.) Juice on Listeria Monocytogenes Scott and Salmonella Typhimurium ATCC 14028 and The Sensory Properties of Raw Shrimps. *International Journal of Food Microbiology* **136**(1) 88-94

[8] Anitha R and Geetha R V 2011. Averrhoa Bilimbi Linn'Nature's Drug Store-A Pharmacological Review. *International Journal of Drug Development and Research* **3**(3) 1-6

[9] Marcin H, Ryszard K, Anna G and Henryk D 2008 Anthocyanins of Fruits and Vegetables-Their Occurrence, Analysis and Role in Human Nutrition. *Vegetable Crops Research Bulletin* **68** 5-22

[10] Glover B J and Cathie M 2012. *Anthocyanins. Current Biology* **22**(5) R147-R150

[11] Stintzing F C and Carle R 2004 Functional Properties of Anthocyanins and Betalains in Plants, Food and in Human Nutrition. *Trends in Food Science and Technology* **15**(1) 19-38

[12] Ananth S, Vivek P, Arumanayagam T and Murugakoothan P 2014 Natural Dye Extract of Lawsonia Inermis Seed as Photo Sensitizer for Titanium Dioxide Based Dye Sensitized Solar Cells. *Spectrochimica Acta Part A: Molecular and Biomolecular Spectroscopy* **128** 420-426

[13] Al-Alwani M A M, Mohamad A B, Kadhum A A H and Ludin N A 2015 Effect of Solvents on the Extraction of Natural Pigments and Adsorption onto TiO2 for Dye-Sensitized Solar Cell Applications. *Spectrochimica Acta Part A: Molecular and Biomolecular Spectroscopy* **138** 130-137

[14] Sivakumar V, Vijaeeswarri J, Anna J L 2011 Effective Natural Dye Extraction from Different Plant Materials using Ultrasound. *Industrial Crops and Products* **33**(1) 116-122

[15] Hernandez M A R, Estevez M, Vargas S, Quintanilla F and Rodriguez R 2011 New Dye-Sensitized Solar Cells Obtained from Extracted Bracts of Bougainvillea Glabra and Spectabilis Betalain Pigments by Different Purification Processes. *International Journal of Molecular Sciences* **12**(9) 5565-5576

[16] Syafinar R, Nair G, Irwanto M, Fareq M, Irwan Y M 2015 Optical Characterization using Nature Based Dye Extracted from Hibiscus's Flower. *Asian Research Publishing Network (ARPN)* **10** 6336-6340

[17] Wu J, Yue G, Xiao Y, Lin J, Huang M, Lan Z, Tang Q, Huang Y, Fan L, Yin S 2013 An Ultraviolet Responsive Hybrid Solar Cell Based on Titania/Poly (3-Hexylthiophene). *Scientific Reports* **3** 1-6

[18] Georg A, Graf W and Wittwer V 1998 Comparison of Electrical Conductivity and Optical Properties of Substoichiometrically and Electrochemically Coloured WOx Films of Different Crystallinity. *Solar Energy Materials and Solar Cells* **51**(3) 353-370

[19] Berggren L, Ederth J and Niklasson G N 2004 Electrical Conductivity as a Function of Temperature in Amorphous Lithium Tungsten Oxide. *Solar Energy Materials and Solar Cells* **84**(1) 329-336

[20] Girtan and Mihaela 2013 On the stability of the electrical and photoelectrical properties of P3HT and P3HT: PCBM blends thin films. *Organic Electronics* **14**(1) 200-205

Development of longitudinally excited CO_2 laser

N S Masroon*[1], M Tanaka[1], M Tei[1], K Uno[2], M Tsuyama[1], H Nakano[1]

[1] Kindai Univ., 3-4-1 Kowakae, Higashiosaka, Osaka 577-8502, Japan,
[2] U. of Yamanashi, 4-4-37 Takeda, Kohu, Yamanashi 400-8510, Japan

Corresponding author: shahiramasroon@gmail.com

Abstract. Simple, compact, and affordable discharged-pumped CO_2 laser controlled by a fast high voltage solid state switch has been developed. In this study, longitudinal excitation scheme has been adapted for simple configuration. In the longitudinal excitation scheme, the discharge is produced along the direction of the laser axis, and the electrodes are well separated with a small discharge cross-section. Triggered spark gap switch is usually used to switch out the high voltage because of simple and low cost. However, the triggered spark gap operates in the arc mode and suffer from recovery problem causing a short life time and low efficiency for high repetition rate operation. As a result, there is now considerable interest in replacing triggered spark gap switch with solid state switches. Solid state switches have significant advantages compared to triggered spark gap switch which include longer service lifetime, low cost and stable high trigger pulse. We have developed simple and low cost fast high voltage solid state switch that consists of series connected-MOSFETs. It has been installed to the longitudinally excited CO_2 laser to realize the gap switch less operation. Characteristics of laser oscillation by varying the discharge length, charging voltage, capacitance and gas pressure have been evaluated. Longer discharge length produce high power of laser oscillation. Optimum charging voltage and gas pressure were existed for longitudinally excited CO_2 laser.

1. Introduction

Nowadays, lasers are applied in many industrial processes. CO_2 laser is a gas laser that produces a strong infrared light with the principle wavelength bands ranging from on 9.2 to $11.4\mu m$. The excitation schemes used for CO_2 lasers are classified into two according to the structure of the laser tube which are longitudinal excitation scheme and transverse excitation scheme. Pulse discharge, DC discharge and RF discharge are used for the pumping of CO_2 laser.

Here, we focus on the longitudinal excitation scheme pumped by a pulse discharge [1]. In the longitudinal excitation scheme, discharge is produce in the direction of the laser axis, and the electrodes are well separated and give small discharge cross-section. The long discharge length provides a high breakdown voltage at low gas pressure. Laser oscillation at low pressure can be allowed by the uniform discharge due to the fast electron drift velocity. Moreover, the discharge uniformity is not affected by residual charge in the longitudinal excitation because a discharge takes place in a discharge tube with a long length [1-5].

Thyratrons or spark gaps switch have been used in many applications as a discharge switch to switch out of several tens of kilovolts. However, the thyratrons which are expensive and have the life time problem whereas the spark gap switches operate in the arc mode and suffer from recovery problem causing a short life time and low efficiency for high repetition rate operation. Therefore, these switches are being phased out.

Solution to those problems, in recent years, improvement of power semiconductor technology contribute to a major advanced towards highly reliable industrial laser systems since solid state switches are maintenance free and have long life time. In this case, spark gap switch can be eliminated entirely by the use of solid state switch by modern semiconductor switches such as IGBTs and MOSFETs.

2. Experimental

2.1 Longitudinal excitation scheme

The longitudinally discharged-pumped CO_2 laser controlled by a fast high voltage solid state switch is schematically shown in Fig. 1, consisting of high voltage DC power supply, diode, charging resistor, capacitor, discharge tube, and fast high voltage solid state switch. A mixed (CO_2: 70% N_2: 14% He: 79%) gas is filled in a 30 cm and 40 cm-long pyrex tubes with an inner and outer diameter of 10 mm and 14 mm, respectively. Metallic electrodes were attached to the both ends of the laser tube. High reflective mirror and ZnSe output coupler with reflectivity of 70% formed a resonator.

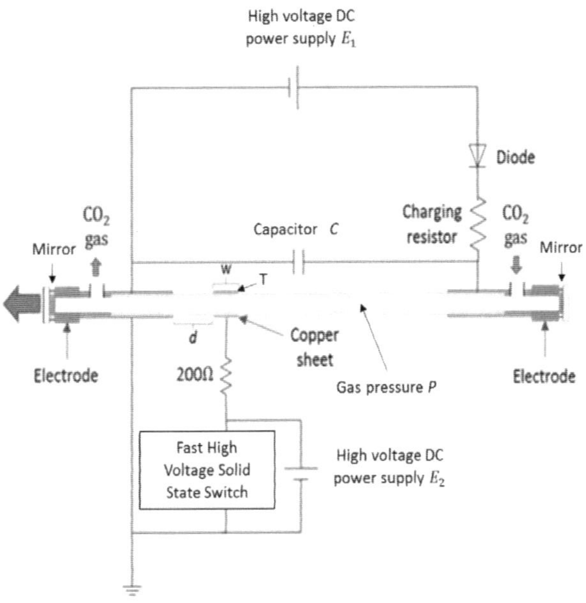

Figure 1. Schematic diagram of longitudinally excited CO_2 laser controlled by a fast high voltage solid state switch

Trigger electrode was made of copper sheet with width of 15 mm and wrapped at the outer wall of the discharge tube. Trigger electrode locates between two main electrodes, is connected with the solid state switch and labelled as T as shown in Fig. 1. The fast high voltage solid state switch indicated in Fig. 1, is operated as a trigger switch. This scheme realizes simple construction of gas laser excitation without discharge switch like a spark gap.

2.2 Fast high voltage solid state switch

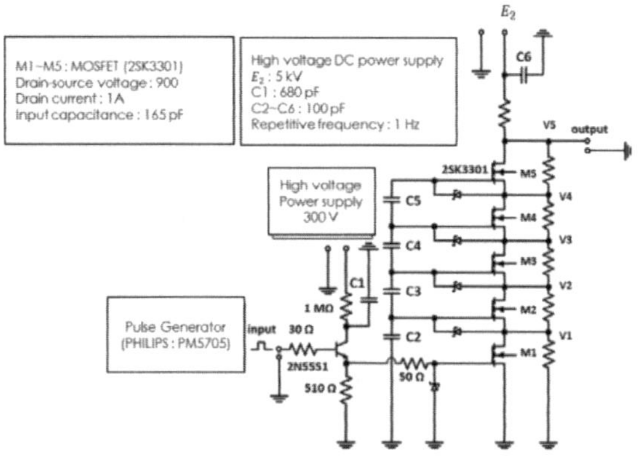

Figure 2. Schematic diagram of fast high voltage solid state switch

The fast high voltage solid state switch that consists of avalanche transistor and series-connected MOSFETs has been developed. An avalanche transistor circuit is used as the gate driver. Connecting several MOSFETs in series allows fast switching time with working-voltages of several kilovolts can be realized [6]. Zener diode between gate and source prevents breaking down of the gate by high voltage pulse [7].

5 kV of DC voltage E_2 is applied from DC power supply to the MOSFETs. The voltage of each MOSFET is divided by the resistances which are connected between drain and source of the MOSFET. At the same time, capacitors are charged. By applying a voltage of 300 V, avalanche breakdown occurs in the avalanche transistor (2N5551). The avalanche transistor produce fast signal in several nanoseconds. This fast pulse charges the input capacitance of the MOSFET (M1) quickly.

When the bottom gate of the M1 turns on, the voltage difference, V1 is generated between ground and the source of M2. V1 is divided by C2 and the input capacitance of M2. The capacitor that connected between gate terminals of every MOSFET is discharge from C2 to C5, and turned on the MOSFETs from M2 to M5 in sequence, finally switch out the high voltage. Fast high voltage solid state switch is triggered as a spiker that produce pre-ionization in laser tube and voltage pulse to initiate the discharge.

This fast high voltage solid state switch capable to deliver fast pulse in 40 ns at 5 kV voltage. This circuit is compact and less expensive since a drive circuit for each gate does not required.

2.3 Laser Oscillations

DC voltage, E_1 was applied to the discharge tube. Capacitor C connected parallel with the discharge tube as shown in Fig. 1 was charged by the current that flows from E_1. At the same time, positive voltage pulse generated by pulse generator was used to trigger the fast high voltage solid state switch. E_2 was set to be 5 kV. When the fast high voltage solid state switch is switched, the discharge is induced in the glass tube producing laser oscillation.

In this experiment, laser output energy was measured with an energy detector (Gentex ED-200), and the voltage waveform was measured with a high voltage probe coupled with an oscilloscope (Tektronik TDS 3054C).

3. Results and Discussion

30 cm and 40 cm-long tubes have needed at least 10 kV and 20 kV charging voltage, respectively to produce laser oscillation. Characteristic of laser output energy was examined under some parameters which are charging voltage, capacitance C, gas pressure and capacitance between trigger and ground electrode to find optimum conditions in order to produce high output energy.

3.1 Relation between capacitance and laser output energy

Laser output energy depends on the capacitance C that connected parallel with the discharge tube. Gas pressure, and distance between trigger and ground electrode were fixed to be 3.0 kPa and 1 cm respectively. In this experiment, capacitance of 570 pF, 700 pF, 900 pF, 1140 pF, 2000 pF and 2700 pF were used. The electrical energy was estimated by using the formula as shown below:

$$E = \frac{1}{2}CV^2 \tag{1}$$

Figure 3 shows dependence of laser energy on electrical energy. Laser energy increases with increasing in electrical energy for both laser tube in 40 cm and 30 cm length. Laser energy with discharge length 40 cm is higher than 30 cm because the high energy was applied. However, for discharge length of 40 cm, laser energy decreases as the electrical energy is more than 400 mJ. When the electrical energy is getting higher, laser oscillation become unstable due to residual charge.

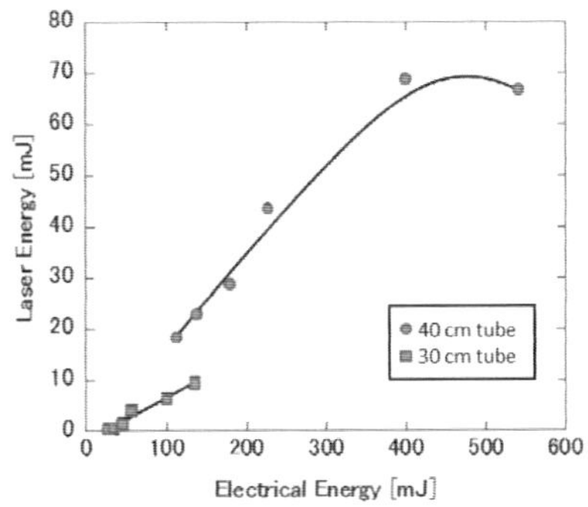

Figure 3. Dependence of the laser output energy on electrical energy

3.2 Relation between pressure and laser output energy

Figure 4 shows the dependence of the laser output energy on gas pressure for 40 cm discharge length. In this case, capacitance C, and distance between trigger and ground electrode were fixed to be 700 pF and 1 cm, respectively. Results shown in Fig. 4 indicates that laser output energy as a function of gas pressure various changing voltage. Optimum gas pressure for laser with discharge length of 40 cm depends on the charging voltage.

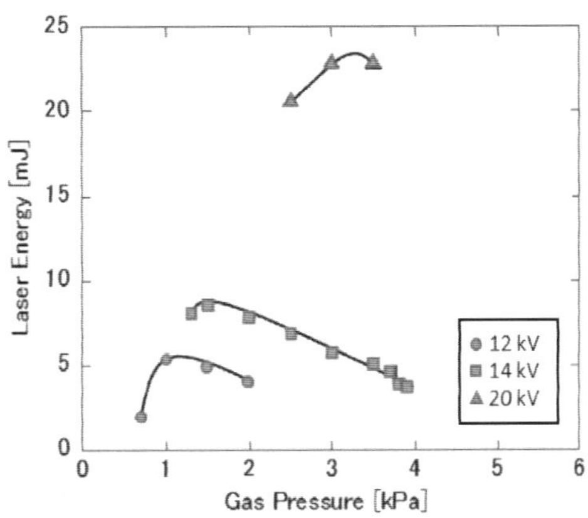

Figure 4. Dependence of the laser output energy on CO_2 gas pressure for different charging voltage

When the gas was supplied at low pressure, the laser output energy become low due to low gas density. Laser output energy have a peak value at gas pressure around 3.0 kPa when 20 kV is being applied, as shown in Fig. 4. Laser energy decreases as the gas pressure flowed more than 3.5 kPa. At 20 kV charging voltage, when the gas pressure is set more than 3.5 kPa, laser energy decreases due to high discharge impedance caused by high gas density. Therefore, the optimization of pressure and charging voltage are required to produce high output energy.

3.3 Relation between capacitance between trigger and ground electrode C_d and laser output energy

In this experiment, 40 cm of discharge length was used. The charging voltage, capacitance C, and gas pressure were fixed to be 20 kV, 700 pF and 3 kPa, respectively, to examine laser energy dependence on capacitance between trigger and ground electrode.

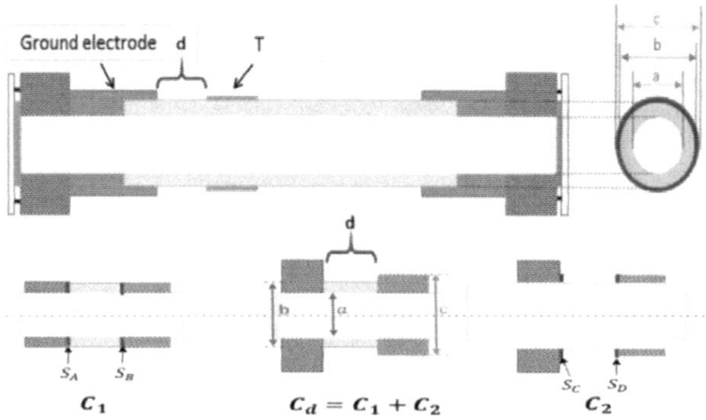

Figure 5. Schematic diagram to calculate capacitance between trigger and ground electrode

To calculate the capacitance between trigger and ground electrode, assuming that C_1 and C_2 were connected in parallel as shown in Fig. 5. The total of capacitance between trigger and ground electrode C_d is equal to the sum of C_1 and C_2. Therefore,

$$C_d = \frac{\varepsilon_0}{d}\varepsilon_r S_A + \frac{\varepsilon_0}{d} S_C \tag{2}$$

where,
a: inner diameter of laser tube
b: outer diameter of laser tube
c: outer diameter of copper sheet
d: distance between trigger and ground electrode
S_A, S_B: cross-sectional area of laser tube
S_C, S_D: cross-sectional area of copper sheet
ε_0: permitivity of free space
ε_r: relative permittivity

Distance between trigger and ground electrode which are 0.5 cm, 1.0 cm, 1.5 cm and 2.0 cm correspond to the 0.65 pF, 0.32 pF, 0.22 pF and 0.16 pF, of capacitances, respectively. As shown in Fig. 6, laser oscillates when the distance between trigger and ground electrode is within 0.5 cm to 2 cm. When the distance is set more than 2 cm (trigger capacitance less than 0.16 pF), the laser oscillation did not occur. As the distance between trigger electrode and ground electrode increases, the charge is not satisfied to trigger the discharge. Hence, optimum trigger capacitance is existed for this laser scheme.

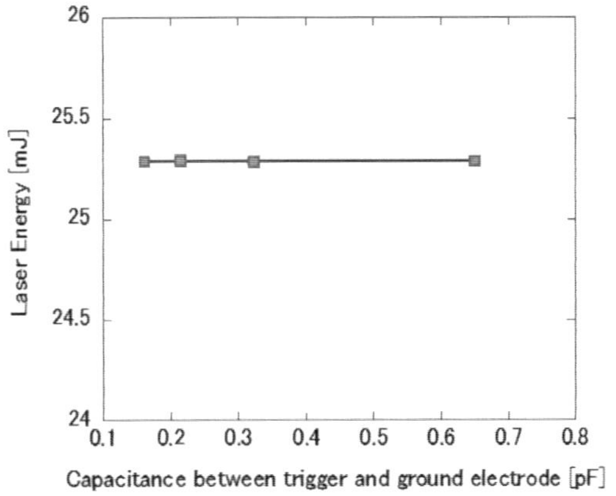

Figure 6. Dependence of laser output energy on capacitance between trigger and ground electrode

4. Conclusion

Simple, compact and affordable longitudinally excited CO_2 laser controlled by fast high voltage solid state switch has been developed. In conclusion, longer discharge length and high voltage produce stable and high power of laser oscillation. Additionally, this laser scheme can successfully make an oscillation at several conditions that optimum electric charge, gas pressure, and capacitance between trigger and electrode. There are various factors that influence the performance of longitudinally excited CO_2 laser need to be concern.

References

[1] Uno K, Akitsu T and Jitsuno T 2012 Longitudinally excited CO2 laser with short laser pulse using direct-drive circuit. *Journal of Engineering and Technology* **2** 103-108

[2] Uno K, Hayashi H, Akitsu T and Jitsuno T, 2013 Fast Discharge Circuit for Longitudinally Excited CO_2 Laser, *Journal of Infrared, Millimeter and Terahertz Waves* **34** 217 -224

[3] Uno K, Hayashi H, Akitsu T and Jitsuno T 2013 Relation between discharge length and laser pulse characteristics in longitudinally excited CO_2 laser. *Journal of Infrared, Millimeter and Terahertz Waves* **34** 225-230

[4] Uno K, Dobashi K, Akitsu T and Jitsuno T 2015 Longitudinally excited CO_2 laser with tail-free short pulse. *Proc. SPIE 9266,* 92661U

[5] Tsushima H, Uno K, Akitsu T and Jituno T 2017 Longitudinally excited N_2 laser with large-diameter discharge tube. *Rev. Sci. Instrum.* **88** 043106

[6] Gester C 1994 Fast high power / high voltage switch using series connected IGBTs with active gate-controlled voltage balancing. *Conf. Proc. Applied Power Electronics Conf.* 469-472

[7] R J Baker and M D Pocha 2005 Nanosecond switching using power MOSFETs. *Rev. Sci. Instrum.* **61** 1182

IOP Conf. Series: Journal of Physics: Conf. Series **1027** (2018) 012011 doi:10.1088/1742-6596/1027/1/012011

Generation of an ultrafast femtosecond soliton fiber laser by carbon nanotube as saturable absorber

M A M Salim[*1,2], H Ahmad[1], S W Harun[3], N Bidin[2] and G Krishnan[2]

[1]Photonics Research Centre, University of Malaya, 50603 Kuala Lumpur, Malaysia
[2]Laser Center, Ibnu Sina Institute for Scientific & Industrial Research, Universiti Teknologi Malaysia, 81310 Johor Bahru, Malaysia
[3]Eletrical Engineering Department, Faculty of Engineering, University of Malaya, Kuala Lumpur, Malaysia

Corresponding Author: aizi@siswa.um.edu.my

Abstract. This paper reports the demonstration of ultrafast fiber laser in a simple erbium-doped fiber (EDF) laser that employed a carbon nanotube (CNT) thin film saturable absorber (SA) to generate a stable soliton pulse. The repetition rate of 10.8 MHz pulse consistently achieved has narrowest pulse width of 640 fs and 1555.78 nm central wavelength for an hour operation in room temperature. This proposed setup has the capability for reliable and stable system features.

1. Introduction

Passively ultrafast femtosecond erbium-doped fiber laser (EDFL) with its uniqueness and compactness has gained a wide attention in pulse fiber laser application. A pulse laser has opened up an extensive variety of laser applications, for instance supercontinuum generation [1, 2], laser marking and cutting [3] and non-linear nonlinear frequency conversion [4, 5]. Pulse laser has been reported to have a low cost deployment and well known for their flexibility design [6, 7].

A number of techniques have been widely reported to generate passive ultrafast femtosecond soliton fiber laser. One of the effective methods is to use an optical material in this case refer to saturable absorber (SAs), where at highest optical intensities the optical loss can be decreased. In the last decade, semiconductor saturable absorber mirror or SESAM has been a dominant SAs technology to produce passively Q-switched pulse [8, 9]. However, there are several drawbacks associated with SESAM as SAs for pulse laser. For example, SESAM requires expensive clean room facilities and have a very complex packaging and fabrication [10, 11]. In addition, they also suffer from the narrowband wavelength range [12]. In recent years, a carbon-based nanomaterials such as carbon nanotubes (CNTs) [12-14] and graphene [15, 16] have been introduced to overcome the limitation of SESAMs in Q-switching technology. CNT-based SAs are still in demand as they have interesting features such as simple fabrication process and ultrafast recovery time that make them as practical SAs for pulse laser application.

In this work, we demonstrated an ultrafast soliton mode locked EDFL by using CNT film as SA. The soliton spectrum was obtained at centered wavelengths of 1555.78 nm. The pulse repetition rates and pulse width obtained from the setup were 10.8 MHz and 640 fs respectively thus demonstrated the CNT characteristic of saturable absorption for EDFL mode locking operation.

2. Experimental Setup

In this experiment, mode locked setup is depicted in figure 1 which comprises of 50 cm of Liekki ER110-4/125 erbium doped fiber with 110 dB/m peak absorption at 1530 nm, a single mode fiber (SMF) and few layer carbon nanotube (CNT) based SA. The erbium-doped fiber was forward pumped by a Oclaro LC96A74P-20R laser diode (LD) with 974 nm wavelength and 600 mW maximum pump power. The LD was coupled to the laser cavity via a 980/1550 nm wavelength division multiplexer (WDM). Then, a polarization insensitive isolator is incorporated into the laser cavity to avoid undesired backward reflection propagation and was attached to polarization controller (PC) and then was attached to few layers CNT, which functions as a SA, was sandwiched between two fiber ferrules. In this experiment, the preparation and fabrication of few layers CNT was detail explained in our previous work [17]. A 10-dB combined bi-conical optical coupler (OC1) was used to tap 10% of total power in-order to monitor the optical spectrum, power and mode locked pulses. 10% output port was afterwards attached to a 3-dB optical coupler (OC2) to enable simultaneous monitoring of pulse traces and wavelength spectrum. 90% output port was connected to 500 m SMF and then attached to WDM to complete cavity ring configuration. During the experiment, one of the output coupler was swapped to an OPM to measure the output power. A YOKOGAWA AQ6373 optical spectrum analyzer (OSA) with 0.02 nm resolutions and a fast Thorlabs D400FC photodetector were connected to YOKOGAWA DLM2054 digital oscilloscope to monitor the pulse train. A 9 kHz -7.8 GHz radio frequency spectrum analyzer (RFSA) model MS2683A by Anritsu is utilized to measure the pulse signal in frequency domain and an autocorrelator model HAC200 (ALNAIR) is used to measure the pulse width of soliton mode locked pulse.

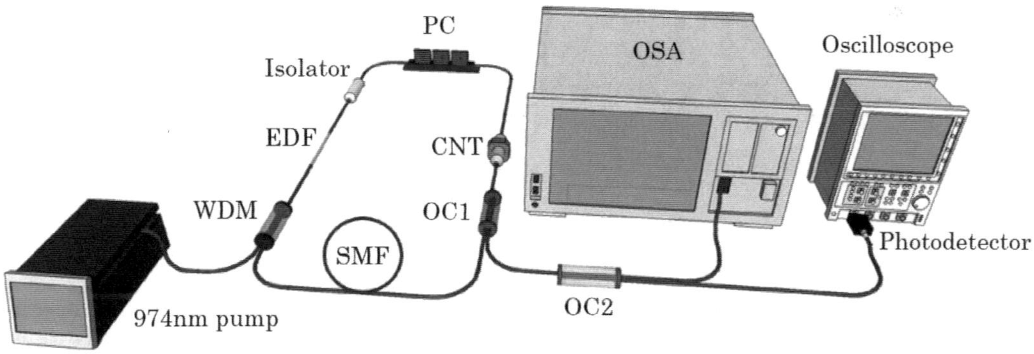

Figure 1. Schematic CNT based mode locked in the EDFL

3. Results and Discussion

In this work the EDFL began its mode locked self-started operation at 80 mW pump power, as given in figure 2(a) obtained from the OSA, with soliton Kelly sidebands spectrum at 1555.78 nm and the output power of -36.74 dBm. Although the pump power was increased, a stable pulse train with constant repetition rates were observed by the oscilloscope as depicted in figure 2 (b) with repetition rates of 10.8 MHz. The mode locked frequency was not relying on the pump power, as a common behavior of a passively Q-switched fiber laser. Furthermore, figure 2 (c) shows the pulse width was measured to be 640

IOP Conf. Series: Journal of Physics: Conf. Series **1027** (2018) 012011 doi:10.1088/1742-6596/1027/1/012011

fs (full width at half maximum) using autocorrelator. Figure 2 (d) records the frequency domain of the pulse train, achieved from a radio frequency spectrum analyzer (RFSA). The spectrums verify 10.8 MHz as the fundamental harmonic frequency with a 38 dB peak-to-pedestal ratio. Furthermore, the wideband of RFSA as given in figure 2 (d) shows no spectral modulation, implying a stable soliton mode locked pulse generated in the system.

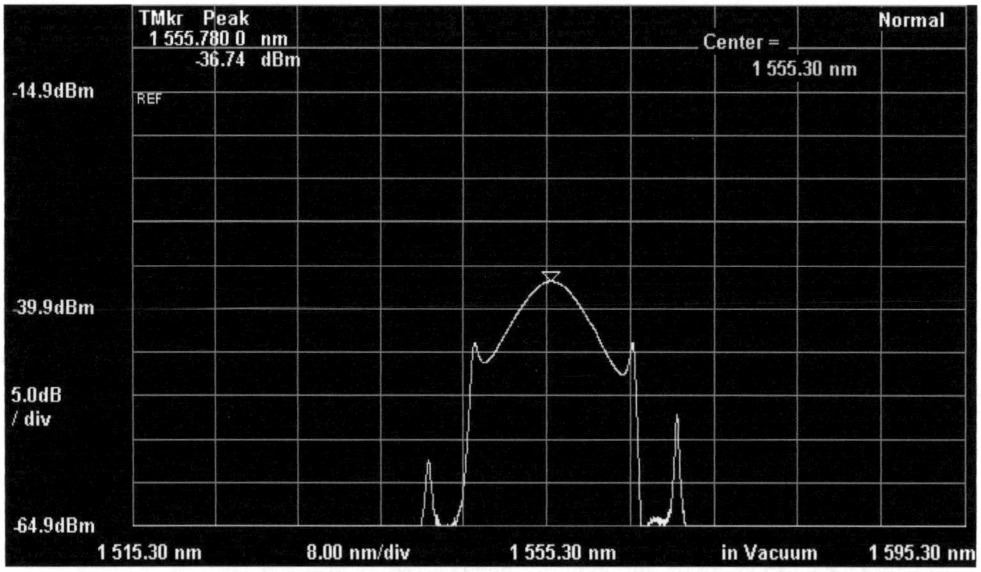

(a)

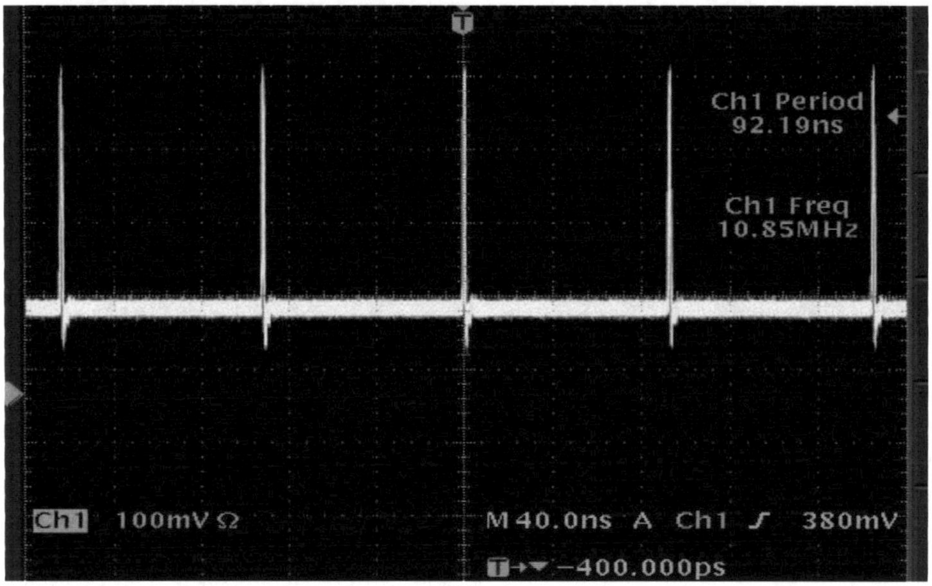

(b)

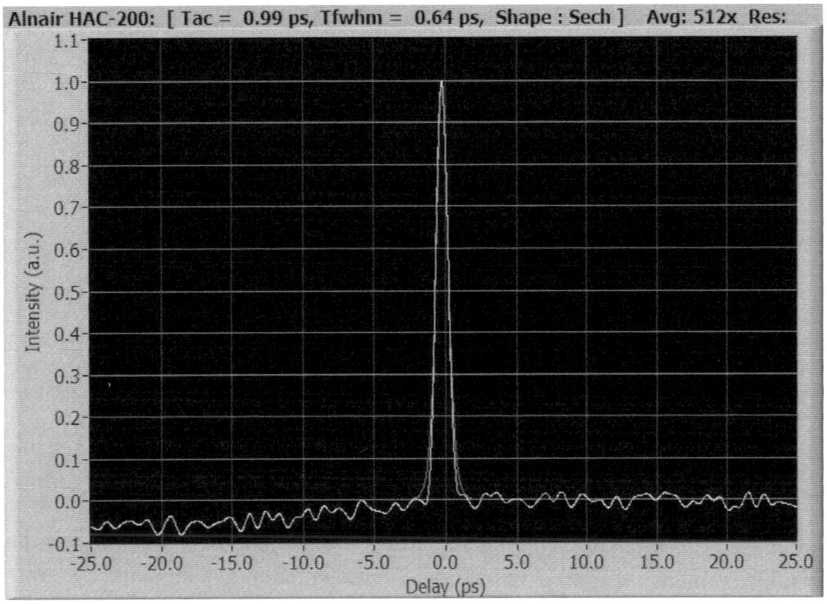

(c)

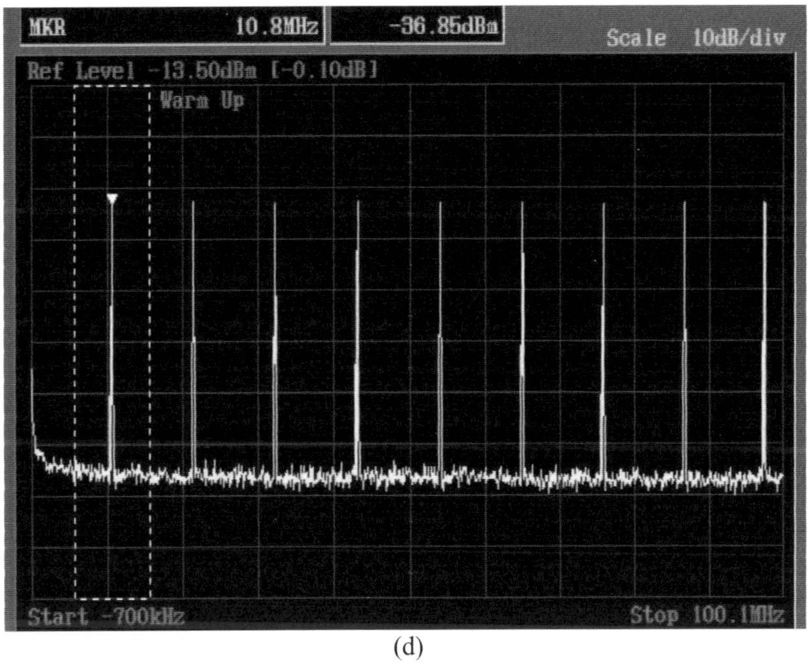

(d)

Figure 2. The soliton mode locked characterization at 80 mW with (a) soliton Kelly sidebands spectrum, (b) pulse duration, (c) pulse width and (d) radio frequency output spectrum.

4. Conclusion

In summary, we have successfully achieved the soliton mode locked based CNT SA. The soliton Kelly sidebands spectrum was obtained at 1555.78 nm wavelength and repetition rate of 10.8 MHz with narrowest pulse width of 640 fs.

Acknowledgement

We would wish to thank the University of Malaya for the research funding, under grant number RU 0016/2014, RP008A-13AET, RG143-12AET and Universiti Teknologi Malaysia under grant number R.J130000.7809.4L671 and Q.J130000.2526.15H47.

Reference

[1] Sukhoivanov I A, Iakushev S O, Shulika O V, Andrade-Lucio J A, Díez A and Andrés M 2014 Supercontinuum generation at 800 nm in all-normal dispersion photonic crystal fiber. *Optics express* **22**(24) 30234-30250

[2] Roy A, Laroche M, Roy P, Leproux P and Auguste J L 2007 Q-switched Yb-doped nonlinear microstructured fiber laser for the emission of broadband spectrum. *Optics letters* **32**(22) 3299-3301

[3] Okamoto Y, Kitada R, Uno Y and Doi H 2008 Cutting of solid type molded composite materials by Q-switched fiber laser with high-performance nozzle. *Journal of Advanced Mechanical Design, Systems, and Manufacturing* **2**(4) 651-660

[4] Shi W, Leigh M, Zong J and Jiang S 2007 Single-frequency terahertz source pumped by Q-switched fiber lasers based on difference-frequency generation in GaSe crystal. *Optics letters* **32**(8) 949-951

[5] Shi W, Leigh M A, Zong J, Yao Z, Nguyen D T, Chavez-Pirson A and Peyghambarian N 2009 High-power all-fiber-based narrow-linewidth single-mode fiber laser pulses in the C-band and frequency conversion to THz generation. *IEEE Journal of Selected Topics in Quantum Electronics* **15**(2) 377-384

[6] Dupriez P, Finot C, Malinowski A, Sahu J K, Nilsson J, Richardson D J and Tropper A C 2006 High-power, high repetition rate picosecond and femtosecond sources based on Yb-doped fiber amplification of VECSELs. *Optics express* **14**(21) 9611-9616

[7] Leigh M, Shi W, Zong J, Wang J, Jiang S and Peyghambarian N 2007 Compact, single-frequency all-fiber Q-switched laser at 1 μm. *Optics letters* **32**(8) 897-899

[8] Spühler G J, Paschotta R, Kullberg M P, Graf M, Moser M, Mix E and Keller U 2001 A passively Q-switched Yb: YAG microchip laser. *Applied Physics B* **72**(3) 285-287

[9] Shi-Xiang X, Wen-Xue L, Qiang H, Hui Z and He-Ping Z 2008 Efficient laser-diode end-pumped passively Q-switched mode-locked Yb: LYSO laser based on SESAM. *Chinese Physics Letters* **25**(2) 548

[10] Ahmad H, Salim M A M, Azzuhri S R, Soltanian M R K and Harun S W 2015 A passively Q-switched ytterbium-doped fiber laser based on a few-layer Bi2Se3 saturable absorber. *Laser Physics* **25**(6) 065102

[11] Ahmad H, Salim M A M, Soltanian M R K, Azzuhri S R and Harun S W 2015 Passively dual-wavelength Q-switched ytterbium doped fiber laser using Selenium Bismuth as saturable absorber. *Journal of Modern Optics* **62**(19) 1550-1554

[12] Schmidt A, Rivier S, Steinmeyer G, Yim J H, Cho W B, Lee S and Díaz F 2008 Passive mode locking of Yb: KLuW using a single-walled carbon nanotube saturable absorber. *Optics letters* **33**(7) 729-731

[13] Gao Y, Zhao T Z, Li C Y, Ge W Q, Wu Q, Shi Z H and Wang Y G 2013 Diode-side-pumped

passively Q-switched Nd: YAG laser at 1123 nm with reflective single walled carbon nanotube saturable absorber. *Optics Communications* **286** 261-264

[14] Ahmad H, Semangun J M, Azzuhri S R, Zulkifli M Z, Awang N A and Harun S W 2015 Passively mode-locked laser using an entirely centred erbium-doped fiber. *Laser Physics* **25**(4) 045105

[15] Liu J, Wu S, Yang Q H and Wang P 2011 Stable nanosecond pulse generation from a graphene-based passively Q-switched Yb-doped fiber laser. *Optics letters* **36**(20) 4008-4010

[16] Zhang L, Fan J T, Wang J H, Hu J M, Lotya M, Wang G Z and Feng Y 2012 Graphene incorporated Q-switching of a polarization-maintaining Yb-doped fiber laser. *Laser Physics Letters* **9**(12) 888

[17] Ismail M A, Harun S W, Zulkepely N R, Nor R M, Ahmad F and Ahmad H 2012 Nanosecond soliton pulse generation by mode-locked erbium-doped fiber laser using single-walled carbon-nanotube-based saturable absorber. *Applied optics* **51**(36) 8621-8624

Libs-PCA based discrimination of Malaysian coins

Auwal Mustapha Imam[1,2,3], M. Safwan Aziz[1,3], Kashif Chaudhary[1,3], Zuhaib Rizvi*[1], Jalil Ali[1,3]

[1]Laser Centre, Ibnu Sina Institute for Scientific and Industrial Research (ISI-SIR), Universiti Teknologi Malaysia (UTM), Johor Bahru, 81300, Malaysia.
[2]Department of Physics with Electronics, Federal University Birnin Kebbi, 862104, Kebbi State, Nigeria.
[3]Physics Department, Faculty of Science, Universiti Teknologi Malaysia (UTM), Johor Bahru, 81300, Malaysia.

Corresponding Author: zuhaib@utm.my

Abstract. The investigations of currency coins dated back to many centuries. Many researchers developed an interest in the investigation of the coin's weight, size, physical feature and elemental composition. Laser-induced breakdown spectroscopy (LIBS) has the novelty of analytical analyses of various samples. It has the ability for the elemental composition determination of samples of solid (including metals), liquid and/or gases. Malaysia as a country uses Ringgit as a currency, among which are coins of 10, 20 and 50 cents. These coins are in series of release from the Malaysian Central Bank from time to time. There are currently in circulation old and new coins of 5, 10, 20 and 50 cents coins. These coins differ in their physical features and are may be different also in their elemental composition. This paper presents the investigation of the differences in elemental composition between the old and new Malaysian coins of 10, 20 and 50 cents. Principal component analysis (PCA) was used to perform the discrimination between the coins from the LIBS spectra.

1. Introduction

Laser-induced breakdown spectroscopy (LIBS), also known as laser-induced plasma spectroscopy (LIPS), is one of the common and widely used analytical techniques based on optical emission spectroscopy [1, 2], it is an excellent, convenient and less-expensive analytical technique [3]. LIBS was first introduced in the 1960s as an elemental analytical technique. [4], and became popular after the development of sophisticated detectors and spectrometers [5]. The technique was first developed for specific applications in the laboratories, but it was available for general applications in 1990s. It was not until 1990s that LIBS found attention in general applications [6]. When a laser beam is focused on the surface of a sample, some amount of the material is ablated, vaporized and ionized to form a plasma plume with a spectral light having a characteristic of the material's elemental composition [7]. The knowledge of the sample ablation, plasma formation and optical emission spectroscopy has brought about advancement in analytical science and techniques, resulting in fast and robust method of identification, classification and discrimination of various samples of a wide variety of materials. This combined technique is known as the laser-induced breakdown spectroscopy (LIBS) [8]. LIBS has been applied successfully for the samples characterization [9], dynamics [10], classification and discrimination of different samples, including soil [11], leaves [12], rocks and soils [13], alloys [14], polymer [15], etc. The type of spectrometer to be used in LIBS elemental analyses requires a balance among the spectral resolution, wavelength, time, limit of detection (LOD) and the dynamic range [16].

LIBS integrated with rain forest (RF), out-of-bag (OOB) estimation, cross-validation (CV), partial squares discriminant analysis (PLS-DA) and support vector machines (SVM) was developed and applied for the discrimination and identification of nine steel grades in [17]. Costa et al [15] proposed a rapid measurement technique using LIBS and classification models such as k-nearest neighbors (KNN) and soft independent modelling of class analogy (SIMCA) to rank polymers. Campanella et al [18] recently used LIBS spectra based on artificial neural networks for the classification of aluminum alloys. The classification was done based on simple emission line ratios. Principal component analysis (PCA) algorithm for feature extraction and multilayer perceptron algorithms and neural networks were applied for classification of laser-induced breakdown spectroscopic data of four proteins: Bovine Serum Albumin, Osteopontin, Leptin and Insulin-like Growth Factor II was studied in [19].

Principal component analysis (PCA) is one of the powerful techniques in chemometrics. PCA is sometimes complemented by an algorithm to get the required classification [20]. LIBS has the capability to distinguish between different materials, even if the difference between the composition elements is minute [21]. Unnikrishnan et al [22] applied 3rd harmonic (355 nm) nanosecond pulse from an Nd:YAG laser to create plasma on the sample surface of plastic and identified the type of the plastic from the plasma emission. Principal Component Analysis (PCA) of the LIBS spectra was employed for the classification of the plastics. Vitkova et al [23] applied the combination of principal components analyses (PCA) and linear discriminant analyses (LDA) on LIBS spectra with multivariate statistical methods for comparative classification of brick samples. LIBS were used to classify the 29 brick samples from 7 different locations. Stand-off and table-top set-ups were used to compare the obtained results. All these studies prove the capability of the LIBS technique for classification of a variety of samples, no matter how little their difference may be.

Investigation of coins has over the years generated interest among many researchers, there is high interest in classification of these coins to originals or copies, old or new. There is challenge in explaining the weight, physical features, composition [24] or even the size of the coins. Other methods such as x-ray fluorescence (XRF) [25, 26], scanning electron microscope (SEM) [27] etc. have been applied for coins analyses. Investigations of the differences in the currency coins allow to obtain information on their elemental composition. LIBS has the ability to perform the differentiation, discrimination and identification of coins. This study is intended to study the differences in physical features and elemental compositions of old and new Malaysian coins (50, 20 and 10 cents) using the LIBS technique.

2. Experimental Setup

The schematic illustration of experimental setup used in this study is presented in Figure 1. A Q-Switched Nd:YAG laser (1064 nm, 6 ns, 650 mJ, 10 Hz) was focused onto the sample surface using a converging lens. Focused laser created a plasma plume on the sample surface, radiations emitted by the excited plasma species were collected by fiber optic cable aligned at 45° to the normal. Radiations were turned into spectrum by a 7-channel spectrometer (OceanOptics LIBS2500+). Gate delay and exposure times were kept constant during LIBS measurements at 3.75 μs and 1 μs respectively. Each sample (old and new coins of 10 cents, 20 cents and 50 cents) was exposed to 5 cleaning shots and 10 measurement shots at 10 different spots on the surface. In this way, 100 LIBS spectra were recorded for each coin.

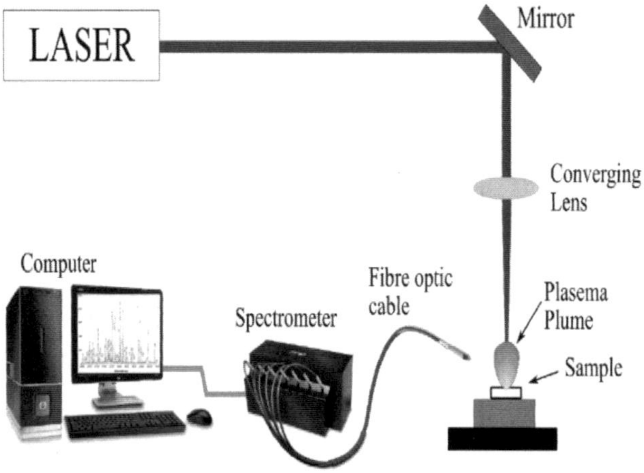

Figure 1. Schematic diagram of the experimental setup.

Datasets to perform Principal Component Analysis (PCA) based discrimination on old and new sets of coins, set of old coins only and set of new coins only. Each set contained equal numbers of spectra for each set of coins. PCA was performed to discriminate among old coins of different values (i.e., 10, 20 and 50 cents).

3. Results and Discussion

Principal component analysis is dimension-reduction tool that reduces a large data to a small set still retaining most of its prominent features. It transforms a number of large number of possibly correlated variables into a smaller number of uncorrelated variables, which are termed as "principal components (PCs)". Most variability of the data is contained by the first principal component; successive components carry as much of the remaining variability of the data. Scores of PCs can be plotted for a three-dimensional view of the data variance.

In this experiment old and new sets of coins 10, 20 and 50 cents of Malaysian currency are subjected to Laser induced breakdown spectroscopy and PCA is performed on their LIBS spectra. PCA brings out discriminatory information in terms of PC scores. In general, first few scores (2 or 3) contain more than 90% variance of a dataset. Therefore, a scatter plot of PC scores gives a pretty clear idea on how distinct are LIBS spectra of materials belonging to different classes.

3.1 Old (second) Series Coins

Looking at the elemental composition of old coins; same elements for 50, 20 and 10 cents (copper-nickel), we assume LIBS will not be able to discriminate between them. Figure 2 shows the images of the old coins while Table 1 illustrates the composition of both the new and old coins.

Figure 2. Old (second) series of Malaysian coins.

Table 1. Details of the elemental compositions of old (second) series and new (third) series of Malaysian coins.

Coin Series	50 Cents	20 Cents	10 Cents
Old	Copper-nickel	Copper-nickel	Copper-nickel
New	Nickel Brass Clad Copper	Nickel Brass	Stainless Steel

3.2 New (third) Series Coins

Looking at the new coins, we assume that the 50 and 20 cents cannot be discriminated, this is because they are made of the almost same material. So, they have the same elemental composition. Likewise, both 50 and 20 cents can be discriminated from the 10 cents, which is made of stainless steel. Generally, all the new coins can be discriminated from the old coins.

Similar separation in the PC scores is observed in Figures 7 and 9 which respectively refer to 20 cents and 50 cents old and new. Whereas, the scatter plot Figures 10 and 11 of old coins of 10, 20 and 50 cents show two clusters.

The cluster of 10 cents is well-separated from the mixed-up cluster of scores of 20 and 50 cents. Clearly, 20 and 50 cents are not discriminable using PCA. Similarly, in Figures 12 and 13, no clear separation between the scores of new coins of 10, 20 and 50 cents are observed making them indistinguishable by PCA. Supervised machine learning techniques may prove helpful in the discrimination of indistinguishable part of this work, which will be the extension of this work.

3.3 LIBS Spectra and PCA

Because of the bulk spectral data we obtained, it was difficult for us to analyze them. Therefore, we employed PCA to help us reduce and compress the data by choosing only principal components (PCs). Even from the LIBS spectra, it will be difficult to analyze the spectral lines.

Figure 3. New series of Malaysian coins

Figure 4 illustrates the spectral lines for the 10 cents old and coins, whereas Figure 5 shows the PCA scores and analyses for the same 10 cents old and new. PCA discriminates very clearly between old and new coins, however could not clearly discriminate between the new coins of 20 and 50 cents. And could not discriminate any from the set of old coins. The reason being the use identical materials for making these coins. The difference in LIBS spectra of two samples arises only if there exists difference between the compositions of the two.

Figure 5, is the scatter plot between PC scores of new and old coins of 10 cents. A clear separation is obtained because of the difference in the materials used to make these coins. Coin of old 10 cents is made of copper-nickel while the coin of new 10 cents is made up of stainless steel. This is the difference in the material that causes a variance in LIBS data and not the shape of the material.

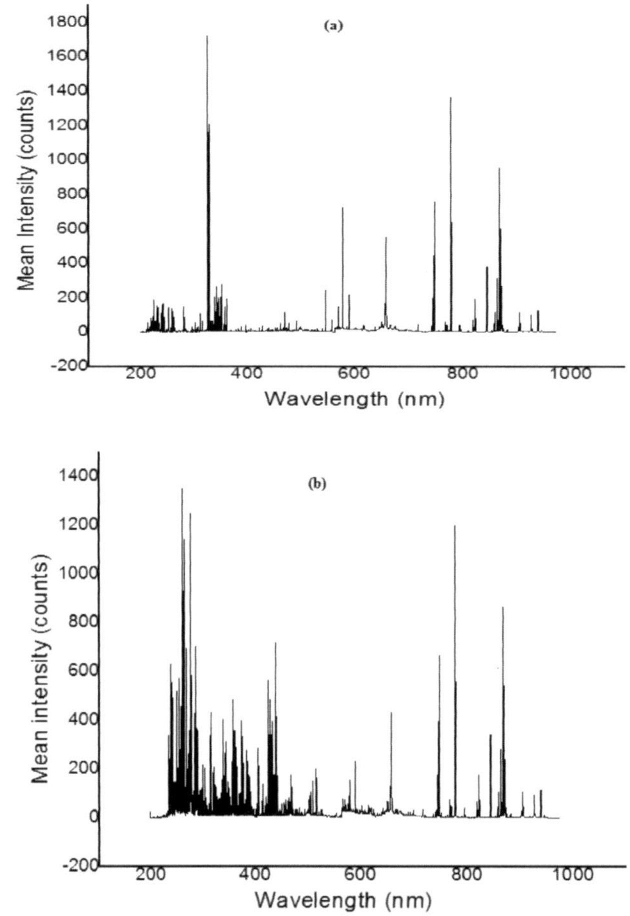

Figure 4. **(a)** 10 cents old and **(b)** 10 cents new spectral lines.

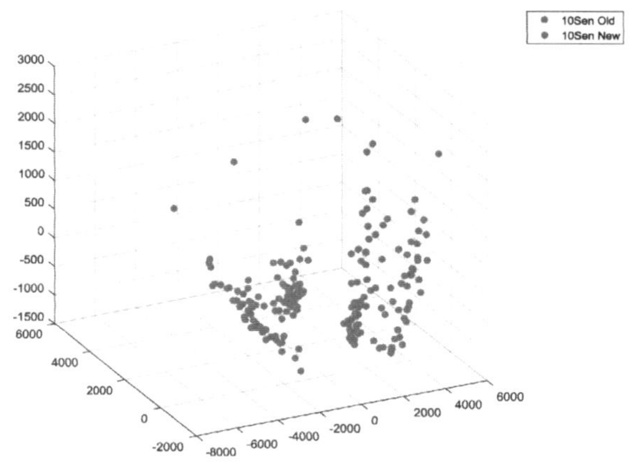

Figure 5. Scatter plot between PC scores of new and old coins of 10 cents.

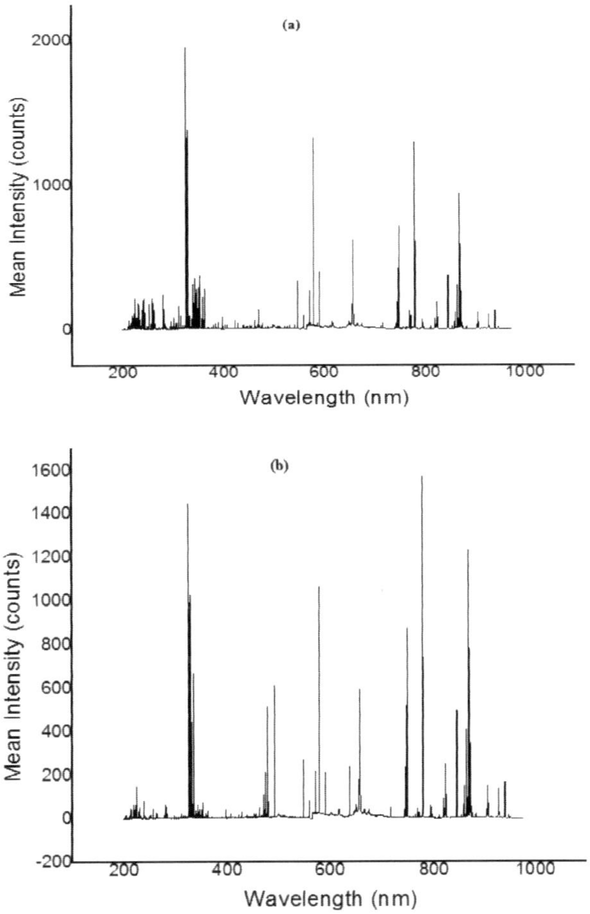

Figure 6. (a) 20 cents old and **(b)** 20 cents new spectral lines.

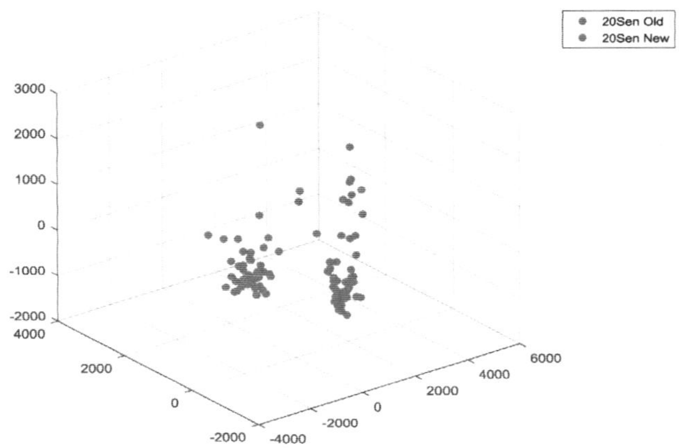

Figure 7. Scatter plot between PC scores of new and old coins of 20 cents.

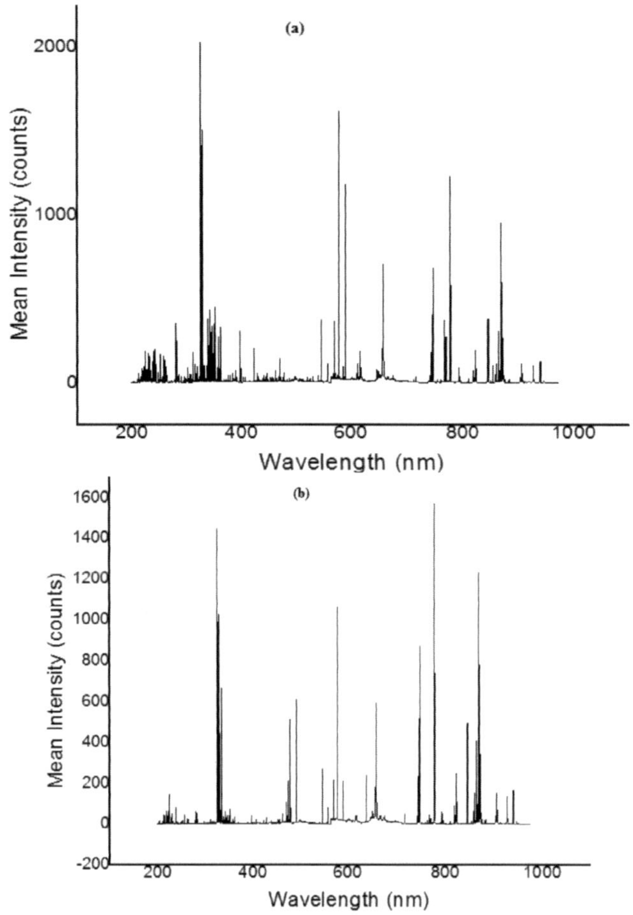

Figure 8. **(a)** 50 cents old and **(b)** 50 cents new spectral lines.

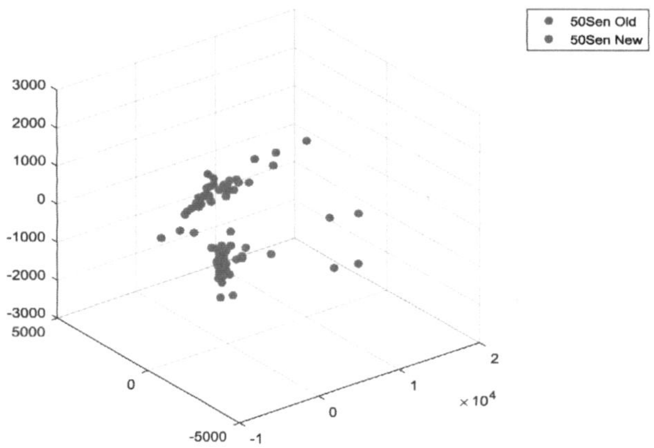

Figure 9. Scatter plot between PC scores of new and old coins of 50 cents.

IOP Conf. Series: Journal of Physics: Conf. Series **1027** (2018) 012012 doi:10.1088/1742-6596/1027/1/012012

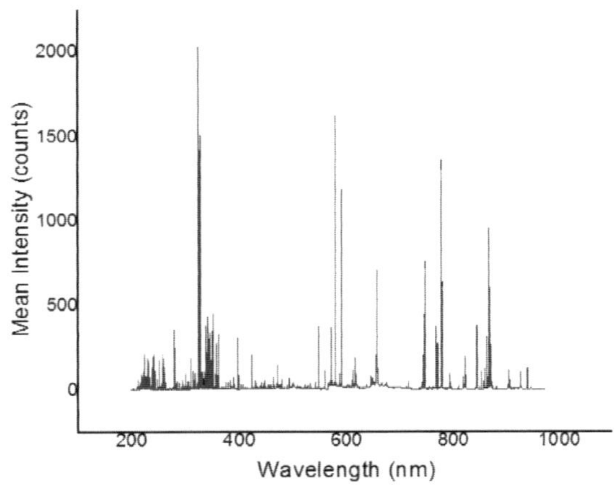

Figure 10. old 50, 20 and 10 cents spectral lines.

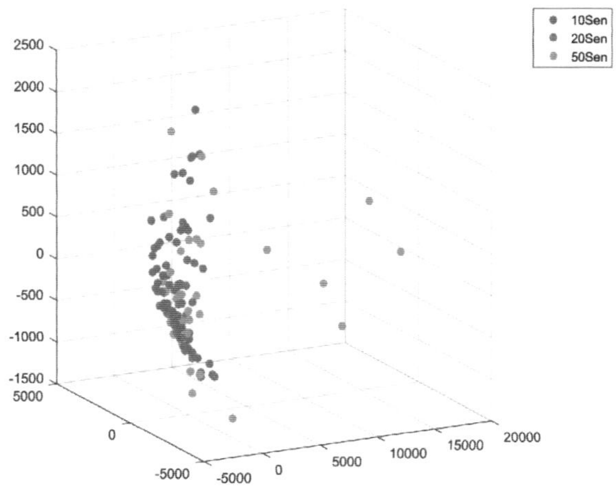

Figure 11. Scatter plot between PC scores of old coins of 50, 20 and 10 cents.

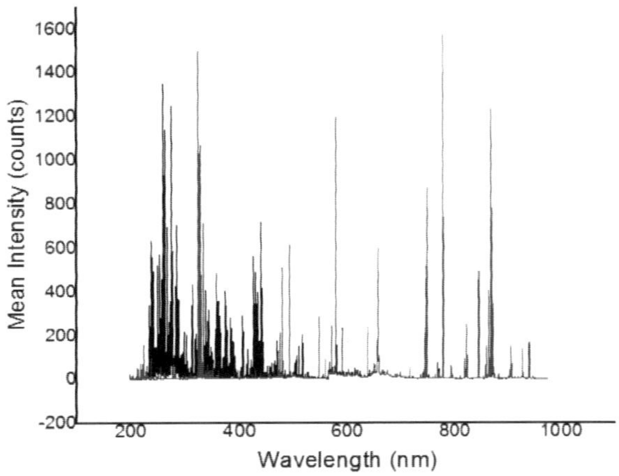

Figure 12. new 50, 20 and 10 cents spectral lines.

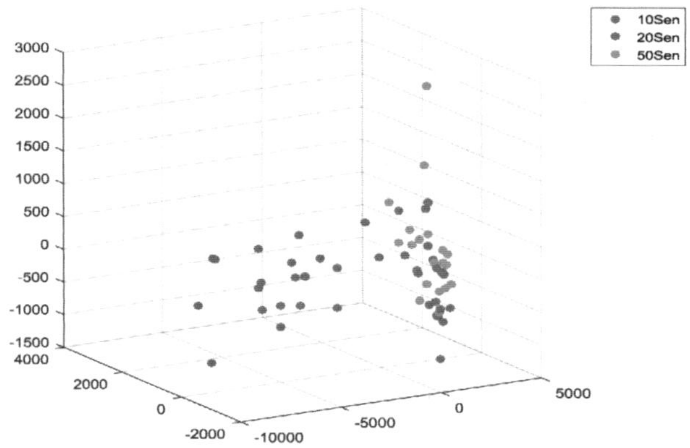

Figure 13. Scatter plot between PC scores of new coins of 50, 20 and 10 cents

4. Conclusion

It is clear, the LIBS technique has the capability to perform discrimination and differentiation of a variety of samples no matter how minute the differences in the elemental composition may be. There was a clear difference between the new coins and the old ones according to the PCA based on LIBS spectra. There was little difference or discrimination for the old coins- this is because they are made of almost same material. For the new 50 and 20 cents, there was no discrimination – this is because the same compositional material was used. LIBS technique was able thus to perform the effective discrimination and classification of coins, as such, it is an effective technique for classification and discrimination of a variety of samples with even minor differences.

Acknowledgement

We would like to thank Ibnu Sina Institute for Scientific and Industrial Research (ISI-SIR), Universiti Teknologi Malaysia (UTM) for providing research facilities. This research work was supported by PDRU (03E22) and Tier 1 (18H67) Grants.

References

[1] De Giacomo A, et. al. 2014 Plasma processes and emission spectra in laser induced plasmas: a point of view. *Spectrochimica Acta Part B: Atomic Spectroscopy* **100** 180-188

[2] Bhatti K, et. al. 2010 Electrons emission from laser induced metallic plasmas. *Vacuum* **84**(7) 980-985

[3] Haider Z, et. al. 2016 Plasma diagnostics and determination of lead in soil and phaleria macrocarpa leaves by ungated laser induced breakdown spectroscopy. *Analytical Letters* **49**(6) 808-817

[4] Runge E, Minck R and Bryan F 1964 Spectrochemical analysis using a pulsed laser source. *Spectrochimica acta,* **20**(4) 733-736

[5] Radziemski L J et. al. 1983 Time-resolved laser-induced breakdown spectrometry of aerosols. *Analytical chemistry* **55**(8) 1246-1252

[6] Radziemski L and Cremers D 2013 A brief history of laser-induced breakdown spectroscopy: from the concept of atoms to LIBS 2012. *Spectrochimica Acta Part B: Atomic Spectroscopy* **87** 3-10

[7] Rakovský J, et. al. 2014 A review of the development of portable laser induced breakdown spectroscopy and its applications. *Spectrochimica Acta Part B: Atomic Spectroscopy* **101** 269-287

[8] Baudelet M and Smith B W 2013 The first years of laser-induced breakdown spectroscopy. *Journal of Analytical Atomic Spectrometry* **28**(5) 624-629
[9] Arab M, et. al. 2015 Characterization of pollution indices in soil surrounding a power plant by laser induced breakdown spectroscopy. *Analytical Letters* **48**(2) 360-370
[10] Chaudhary K, et. al.2015 Laser-Induced Graphite Plasma Kinetic Spectroscopy under Different Ambient Pressures. *Chinese Physics Letters* **32**(4) 043201
[11] Villas-Boas P R, et. al. 2016 Laser-induced breakdown spectroscopy to determine soil texture: A fast analytical technique. *Geoderma* **263** 195-202
[12] Sankaran S, Ehsani R and Morganc K T 2015 Detection of Anomalies in Citrus Leaves Using Laser-Induced Breakdown Spectroscopy (LIBS). *Applied spectroscopy* **69**(8) 913-919
[13] Guang Y, et. al. 2015 Rock and soil classification using PLS-DA and SVM combined with a laser-induced breakdown spectroscopy Library. *Plasma Science and Technology* **17**(8) 656
[14] Castro J P and Pereira-Filho E R 2016 Twelve different types of data normalization for the proposition of classification, univariate and multivariate regression models for the direct analyses of alloys by laser-induced breakdown spectroscopy (LIBS). *Journal of Analytical Atomic Spectrometry* **31**(10) 2005-2014
[15] Costa V C, et. al. 2017 Identification and classification of polymer e-waste using laser-induced breakdown spectroscopy (LIBS) and chemometric tools. *Polymer Testing* **59** 390-395
[16] Arab M, et. al. 2014 Comparison study of two commercial spectrometers for heavy metal analysis of laser induced breakdown spectroscopy (LIBS). *Photonic Sensors* **4**(1) 63-69
[17] Zhang T, et. al. 2016 Classification of steel samples by laser-induced breakdown spectroscopy and random forest. *Chemometrics and Intelligent Laboratory Systems* **157** 196-201
[18] Campanella B, et. al. 2017 Classification of wrought aluminum alloys by Artificial Neural Networks evaluation of Laser Induced Breakdown Spectroscopy spectra from aluminum scrap samples. *Spectrochimica Acta Part-B Atomic Spectrocopy* **134** 52-57
[19] Pokrajac D, et. al. 2010 Performance of multilayer perceptrons for classification of LIBS protein spectra. In *Neural Network Applications in Electrical Engineering (NEUREL), 2010 10th Symposium* 171-174
[20] Xia H and Bakker M 2014 Reliable classification of moving waste materials with LIBS in concrete recycling. *Talanta* **120** 239-247
[21] Chaudhary K, Rizvi S Z H and Ali J 2016 Laser-Induced Plasma and its Applications. In *Plasma Science and Technology-Progress in Physical States and Chemical Reactions.* InTech
[22] Unnikrishnan V, et. al. 2013 Analytical predictive capabilities of laser induced breakdown spectroscopy (LIBS) with principal component analysis (PCA) for plastic classification. *RSC Advances* **3**(48) 25872-25880
[23] Vítková G, et. al. 2014 Comparative study on fast classification of brick samples by combination of principal component analysis and linear discriminant analysis using stand-off and table-top laser-induced breakdown spectroscopy. *Spectrochimica Acta Part B: Atomic Spectroscopy* **101** 191-199
[24] Italiano A, et. al. 2013 A comparative analysis of old and recent Ag coins by XRF methodology. In *3rd Workshop-Plasmi, Sorgenti, Biofisica ed Applicazioni*
[25] Torrisi L, et. al. 2013 Silver coins analyses by X-ray fluorescence methods. *Journal of X-ray science and technology* **21**(3) 381-390
[26] Linke R, et. al. 2003 Determination of the provenance of medieval silver coins: potential and limitations of X-ray analysis using photons, electrons or protons. *X-Ray Spectrometry* **32**(5) 373-380
[27] Torrisi L, et. al. 2009 Laser and electron beams physical analyses applied to the comparison between two silver tetradrachm greek coins. *The European Physical Journal D-Atomic, Molecular, Optical and Plasma Physics* **54**(2) 225-232

Determination of calcium to phosphate elemental ratio in natural hydroxypatite using LIBS

Elham Mazalan[*1], Kashif Chaudhary[1,2], Zuhaib Haider[2], Siti Fatimah Abd Hadi[1] and Jalil Ali[1,2]

[1]Department of Physics, Faculty of Science, Universiti Teknologi Malaysia, Johor Bahru 81310, Malaysia

[2]Laser Centre, Ibnu Sina Institute for Scientific and Industrial Research, Universiti Teknologi Malaysia, Johor Bahru 81310, Malaysia

Corresponding Author: elham.mazalan@gmail.com

Abstract. In this work, laser induced breakdown spectroscopy (LIBS) is employed for determination of calcium to phosphate (Ca/P) ratio in hydroxyapatite (HA) extracted from animal bones. HA was extracted from the biological bones of lamb, bovine, and fish separately. Fresh bones were boiled in deuterium-depleted water (DDW) for 4 hours and immersed it in acetone for 15 hours followed by drying process in the oven. Dried bones were heated at 600 ^{O}C and crushed into powder form which was then mixed with potassium bromide (KBr) in a known quantity. The mixture for each sample was palletized into pellets of identical mass using a hydraulic press. The palletized samples of HA extracted from the bones were ablated with 10 ns Q-switched Nd-YAG laser pulses having an energy of 320 mJ/pulse. The optical emissions produced from the laser induced plasma were recorded. Ca and P emission lines are identified and the Ca/P is calculated. Ca/P ratio by LIBS analysis has been verified by EDX measurements and are found in good agreement.

1. Introduction

LIBS is based on the generation of laser induced breakdown plasma on a sample by focusing a high power pulsed laser beam onto the sample. It has many advantages, which include no requirement of pre-preparation of samples, rapid and low cost analysis [1,2]. From these advantages, LIBS is expected to be the analysis tool in the field of environmental monitoring such as for monitoring the metal element in the polluted soils [3] and analysis of geological materials containing uranium [4]. The LIBS technique also could give information on the different application process such as pulse laser deposition [5] or direct induced plasma [6] by analyzed the properties of the plasma formed. The main advantage of LIBS is that, it is quasi non-destructive, has high throughput and does multi-elemental analysis. However, LIBS is inferior to the conventional element analysis methods such as atomic absorption spectroscopy (AAS), induced coupled plasma-optical emission spectroscopy (ICP-OES), x-ray fluorescence analysis (XRF), and energy-dispersive spectroscopy (EDX) in sensitivity and accuracy.

Hydroxyapatite (HA) is an important bio-medical compound, especially for the preparation of bone cement. The matching Ca/P ratio in HA with human bones provides bonding to generate and fill up missing part of the broken bone without intermediate connective tissue formation. The Ca/P ratio in HA thus is a vital indicator of its compatibility with human bones as bone cement [7]. Studies show that human bone samples and animal meat can be studied using a LIBS technique [8]. The extraction of natural HA from bovine femur bones performed via calcination method at various temperatures affect the composition, crystallinity, and crystallizes size [9]. Hosseinzadeh et al. [10] performed an experiment to derive natural HA from bovine bones with thermal decomposition method by different sample preparation approach. The raw bone is washed and crushed to fine powder. The bone powder was immersed in acetone, then dried in the oven. They mixed the dried bone powder with liquid and pressed before sintering in the furnace. The optimum temperature of 750 °C gives the particle size in the range 420–500 μm.

Another study investigated natural hydroxyapatite and determine the possibility of hydroxyapatite obtained from bovine cortical bone ashes at different thermal conditions [11]. Different sample preparation was used. The bone was cleaned using direct flame thus the organic components were burned. The black powder (bone ash) was placed in an air furnace to remove the remaining char due to the burn. The sample was cooled down inside the furnace until the black ash turned into white granular powder. FTIR analysis showed the existence of organic species and also comprehend the degree of probable dehydroxylation of HA. This paper is focused to determine Ca/P ratio in HA material extracted from animal bones. EDX analysis is performed on an extracted sample of HA from animal bones, (lamb, bovine, and fish) to compare and validate LIBS results.

2. Experimental Setup

This work consists of two main stages, extraction of HA from animal bones and LIBS analysis of extracted HA samples. To extract HA, the fresh bovine, lamb, and fish bones are collected from the local market in Johor Bahru. As shown in Figure 1 (a) bone samples are cut into small pieces. The attached meat, bone marrow and other soft tissues with the bones are removed using a knife. After that, bone samples are boiled in Deuterium-depleted water (DDW) for 4 hours for defatting and easier removal of soft tissues and fats from bones.

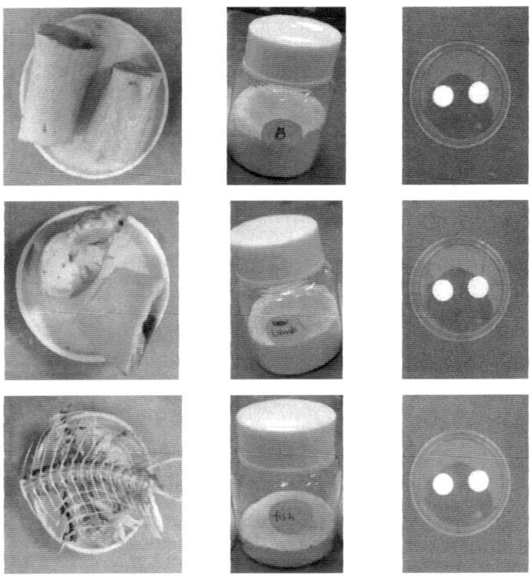

(a) (b) (c)

Figure 1. Bone samples bovine (top), lamb (middle), and fish (bottom). Raw bone sample from market (a), calcined bone sample (b), and pallets of calcined sample (c).

Bone samples are then immersed in acetone for 15 hours for deprotonation followed by drying in oven at 80 °C for 17 hours and then at 600 °C in a furnace for 2 hours. In the final stage, bone samples are crushed into powder using mortar and pestle and calcined at 1000 °C for 2 hours to acquire hydroxyapatite (HA). The calcined samples of bones are shown in Figure 1 (b). The final powder form of the calcined bone samples of all three different types are palletized by mixing a calcined powder sample with KBr by weight percentage 1:10 grams respectively. HA in final pallets form are shown in Figure 1(c).

The experimental setup for real-time LIBS analysis is shown in Figure 2. The first harmonic beam (1064 nm) of 320 mJ energy from Nd:YAG laser was focused onto the sample. An optical fiber cable connected to Maya2000 spectrometer was placed in front of the sample to collect the plasma radiation with an optical response from 200 nm to 650 nm. The spectrometer was connected to the computer and measurement was monitored and recorded using SpectraSuite software. Laser pulse energy was constantly examined by deflecting 6% of the beam to an energy meter through the quartz beam splitter. Each sample is exposed at 10 distinct spots on the surface to 10 laser shots at each spot. For each laser shot an emission spectrum from the plasma is recorded on the computer which were then averaged out. Ca/P ratio is calculated using peak intensities of Ca and P emission lines for all possible combinations of lines and the best one was selected. EDX analysis of the extracted HA from different sources is also performed to calculate the Ca to P ratio. Ca/P ratio calculated by LIBS is with those obtained from EDX.

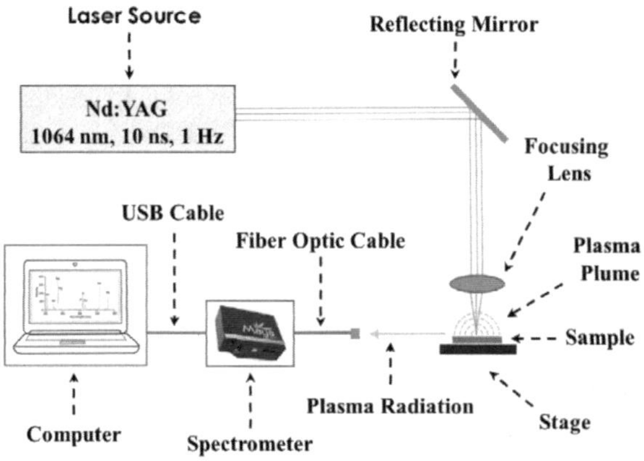

Figure 2. Block experimental set up for LIBS measurement.

3. Results and Discussion

3.1 Element Identification

Figure 3 shows the LIBS spectrum of HA samples for bovine, lamb, and fish samples. Identification of lines is done by matching spectral lines with the NIST atomic spectral database [12]. Identified Ca and P species are listed in the Table 1. Ca and P atomic spectral lines at 442.54 nm and 534.59 nm, respectively, are selected for determining the Ca/P in extracting HA bone samples. Extracted HA from lamb sample has stronger emissions than bovine and fish samples. Each HA sample shows identical spectra of Ca and P.

3.2 LIBS Analysis

Figure 4 and Figure 5 show spectral lines of Ca (Ca I 442.54 nm) and P (P I 534.59 nm) for bovine, lamb, and fish samples. The Ca/P is calculated for each extracted HA bone sample and listed in Table 2. Ca/P of lamb samples has the highest ratio due to the high Ca content.

Table 1. Element identification of lines as comparison with NIST atomic spectral database.

Element	Species (I – Neutral, II – 1st Ionic)	Wavelength, nm
Ca	I	442.54
		422.67
		443.50
		445.48
		610.27
		612.22
		616.22
	II	315.89
		317.93
		370.60
		373.69
		393.37
		396.85
P	I	253.40
		534.59

Table 2. Ca/P ratio for bovine, lamb, and fish bone based on LIBS analysis.

Sample	**Intensity**		**Ca/P ratio**
	Ca (442.54 nm)	**P (534.59 nm)**	
Bovine	2860.80	1493.43	1.92
Lamb	3657.24	1857.63	1.97
Fish	1881.74	1013.49	1.86

IOP Conf. Series: Journal of Physics: Conf. Series **1027** (2018) 012013 doi:10.1088/1742-6596/1027/1/012013

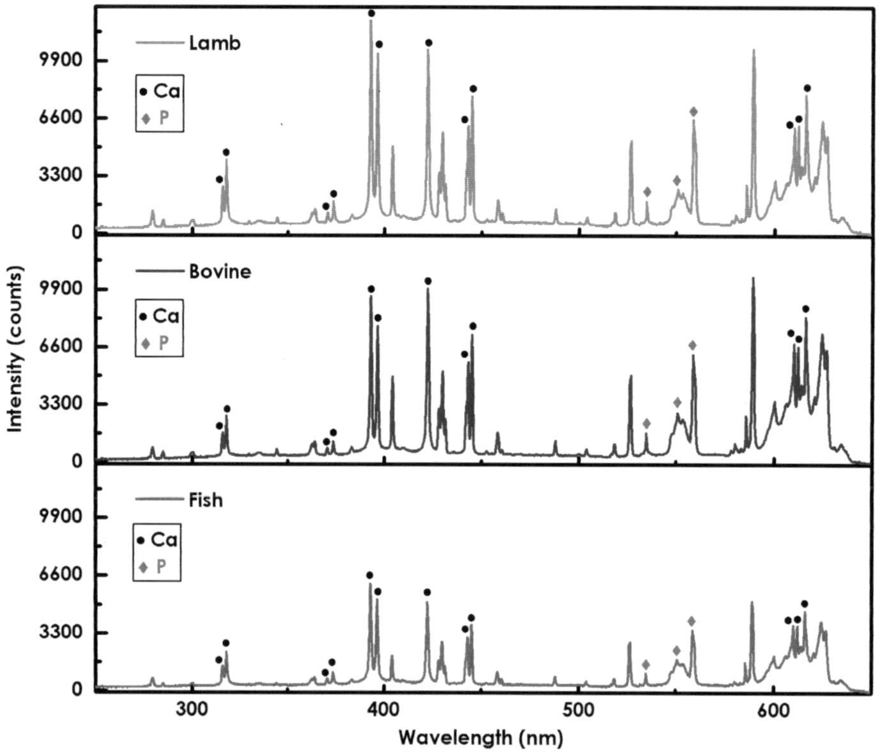

Figure 3. Optical emission spectra of extracted bovine, lamb, and fish samples.

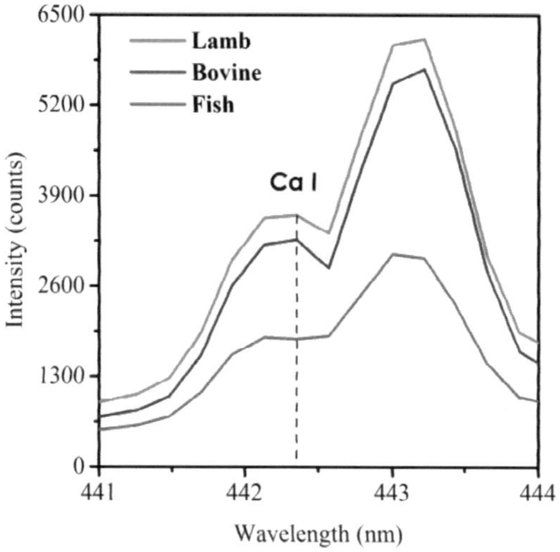

Figure 4. Ca emission spectra of extracted bovine, lamb, and fish samples at 442.54 nm.

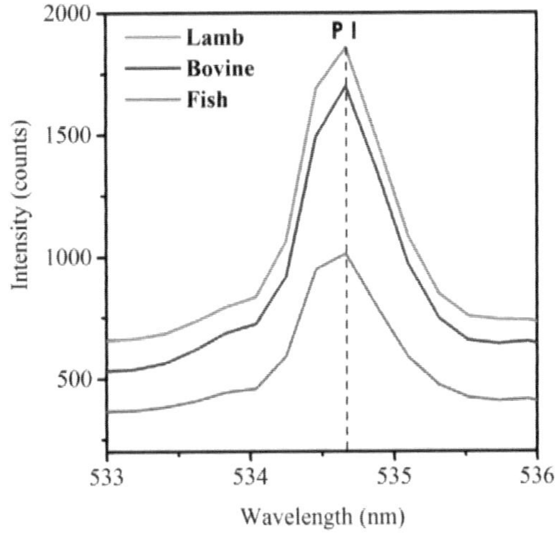

Figure 5. P emission spectra of extracted bovine, lamb, and fish samples at 534.59 nm.

3.3 EDX Analysis

Figure 6 shows the typical EDX spectrum of the extracted HA fish bone samples giving weight percent of Ca and P among other element. The values of Ca/P ratio based on EDX information on extracted HA samples from bovine, lamb and fish bones are given in table 3. The EDX detected carbon (C), Magnesium (Mg), and Oxygen (O) elements, where are Ca and P are present in dominant quantities. All HA bone samples show a higher ratio than the stoichiometric value of HA, which is 1.67 [13].

Table 3. Ca/P ratio for bovine, lamb, and fish bone based on EDX analysis.

Sample	Average Ca wt.%	Average P wt.%	Ca/P ratio
Bovine	39.98	16.62	1.86
Lamb	42.28	17.98	1.82
Fish	42.62	17.70	1.88

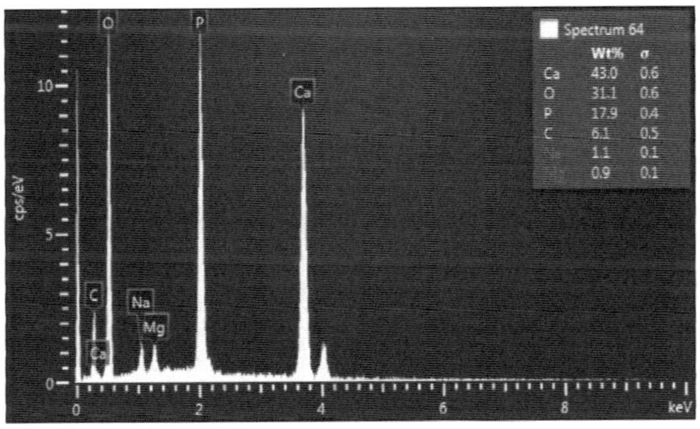

Figure 6. Typical EDX spectra of HA extracted from Fish sample.

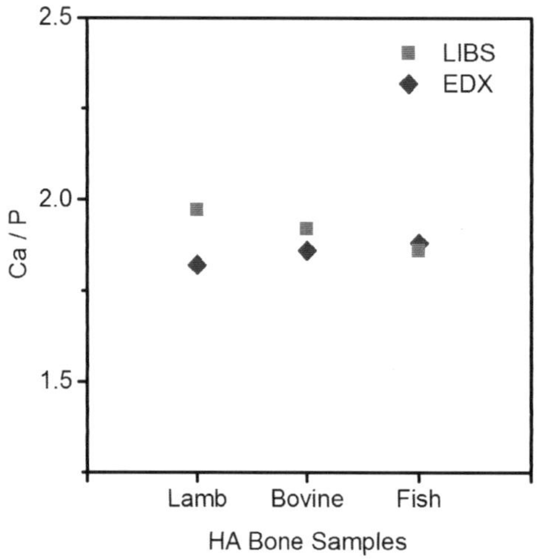

Figure 7. Comparison of LIBS and EDX analysis for lamb, bovine and fish HA bone samples.

3.4 LIBS and EDX Analysis Comparison

The average Ca/P ratio obtained from LIBS analysis were compared with EDX measurements. Figure 7 shows the graph Ca/P ratio for both techniques for each HA bone sample. These values are within the accepted range (1.86-1.97) for bovine, lamb and fish HA bone.

4. Conclusion

In this work, Ca/P ratio in HA extracted from animal sources is determined using LIBS. HA is extracted by the calcination process from bovine, lamb and fish bones. Extracted HA from each animal source was palletized in KBr matrix for LIBS analysis. The pelleted samples of HA were ablated by Q-switched Nd:YAG laser and optical emission in form of spectra are recorded using the optical spectrometer. To verify the LIBS results, EDX analysis of the extracted HA samples has been performed. Ca/P ratios were estimated via the intensity ratio of the selected pair Ca and P lines from the emission spectra. LIBS results were found in good agreement with EDX results. It can therefore be concluded that, LIBS can be used effectively to estimate the Ca/P ratios in HA samples.

Acknowledgement

We would like to thank Ibnu Sina Institute for Scientific and Industrial Research (ISI-SIR), Universiti Teknologi Malaysia (UTM) for providing research facilities. This research work has supported by PDRU (03E22) and Tier 1 (16H42) Grants.

References

[1] H Lin, X Lou, W Zhong and S He 2015 Continuous monitoring of elemental mercury employing low-cost multimode diode lasers. *Meas. Sci. Technol.* **26**(8) 85501

[2] K Chaudhary, S Z H Rizvi and J Ali 2016 Laser-Induced Plasma and its Applications. In *Plasma Science and Technology - Progress in Physical States and Chemical Reactions*, T. B. T.-P. S. and T.-P. in P. S. and C. R. Mieno, Ed. Rijeka. InTech 259–291

[3] M Dell'Aglio, R Gaudiuso, G S Senesi, A De Giacomo, C Zaccone, T M Miano and O De Pascale 2011 Monitoring of Cr, Cu, Pb, V and Zn in polluted soils by laser induced breakdown spectroscopy (LIBS). *J. Environ. Monit.* **13**(5) 1422–1426

[4] J E Barefield, E J Judge, K R Campbell, J P Colgan, D P Kilcrease, H M Johns, R C Wiens, R E McInroy, R K Martinez and S M Clegg 2016 Analysis of geological materials containing uranium using laser-induced breakdown spectroscopy (LIBS). *Spectrochim. Acta - Part B At. Spectrosc.* **120** 1–8

[5] K T Chaudhary, R Qindeel, M S Hussain, J Ali and P P Yupapin 2011 Graphite Thin Film Deposition using Laser Induced Plasma. In *Procedia Engineering* **8** 423–427
[6] K A Bhatti, M Khaleeq-ur-Rahman, M S Rafique, K T Chaudhary and A Latif 2010 Electrons emission from laser induced metallic plasmas. *Vacuum* **84**(7) 980–985
[7] A Niakan, S Ramesh, C Y Tan, J Purbolaksono, H Chandran and W D Teng 2015 Effect of annealing treatment on the characteristics of bovine bone. *J. Ceram. Process. Res.* **16**(2) 223–226
[8] S Moncayo, S Manzoor, T Ugidos, F Navarro-Villoslada and J O Caceres 2014 Discrimination of human bodies from bones and teeth remains by Laser Induced Breakdown Spectroscopy and Neural Networks. *Spectrochim. Acta - Part B At. Spectrosc.* **101** 21–25
[9] W Khoo, F M Nor, H Ardhyananta and D Kurniawan 2015 Preparation of Natural Hydroxyapatite from Bovine Femur Bones Using Calcination at Various Temperatures. *Procedia Manuf.* **2** 196–201
[10] E Hosseinzadeh, M Davarpanah, N H Nemati and S A Tavakoli 2014 Fabrication of a hard tissue replacement using natural hydroxyapatite derived from bovine bones by thermal decomposition method. *Int. J. Organ Transplant. Med.* **5**(1) 23–31
[11] N A M Barakat, K A Khalil, F A Sheikh, A M Omran, B Gaihre, S M Khil and H Y Kim 2008 Physiochemical characterizations of hydroxyapatite extracted from bovine bones by three different methods: Extraction of biologically desirable HAp. *Mater. Sci. Eng. C* **28**(8) 1381–1387
[12] A Kramida, Y Ralchenko and J Reader. NIST Atomic Spectra Database (version 5.4)
[13] S V Dorozhkin 2010 Bioceramics of calcium orthophosphates. *Biomaterials* **31**(7) 1465–1485

Two steps hydrothermal growth and characterisations of $BaTiO_3$ films composed of nanowires

Che Zaheerah Najeehah Che Mohd Zawawi*[1], Shahril Salleh[2], Lee Oon Jew[1] Kashif Tufail Chaudhary[2], Mohamad Helmi[2], Muhammad Safwan Aziz[2], Zuhaib Haider[2], Jalil Ali[2]

[1] School of Ocean Engineering, Universiti Malaysia Terengganu (UMT), 21030 Kuala Terengganu, Terengganu, Malaysia.
[2] Faculty of Science, Universiti Teknologi Malaysia (UTM), 81310 Skudai, Johor Bahru, Johor, Malaysia.

Corresponding author: cz.najeehah@utm.my

Abstract. Barium titanate ($BaTiO_3$) films composed of nanowires have gained considerable research interest due to their lead-free composition and strong energy conversion efficiency. $BaTiO_3$ films can be developed with a simple two steps hydrothermal reactions, which are low cost effective. In this research, $BaTiO_3$ films were fabricated on titanium foil through two steps hydrothermal method namely, the growth of TiO_2 and followed by $BaTiO_3$ films. The structural evolutions and the dielectric properties of the films were investigated as well. The structural evolutions of titanium dioxide (TiO_2) and $BaTiO_3$ nanowires were characterized using X-ray diffraction and scanning electron microscopy. First step of hydrothermal reaction, TiO_2 nanowires were prepared in varied temperatures of 160 ^{0}C, 200 ^{0}C and 250 ^{0}C respectively. Second step of hydrothermal reaction was performed to produce a layer of $BaTiO_3$ films.

1. Introduction

Over recent years, ferroelectric materials have been applied to a large scale of sensors [1-3], actuators [4-6] and energy harvesting systems [7-9] due to their strong electromechanical coupling. Ferroelectric nanomaterials such as thin films [10-13], nanowires [14,15], nanorods[16,17], nanotubes [18,19] and nanocubes [20] have been gaining large attention in nanotechnology. Among these nanostructures, one-dimensional (1D) nanowires are having the smallest dimension that can be employed for efficient transport of optical excitations and electron [21-23]. Besides, it is also expected to show high energy conversion efficiency. Additionally, the high sensitivity and random mechanical disturbances make these nanowires potential candidate for harvesting energy from environment and sensing devices at the nanoscale [24]. In order to directly utilize the ferroelectric arrays in energy harvesting and sensors, the vertically aligned nanowires are typically grown on conductive substrates, which directly act as electrodes for the application for measurement of electric fields [25].

Among ferroelectric materials, barium titanate ($BaTiO_3$) has attracted attention because of having strong ferroelectric properties and thus it is environmental friendly. $BaTiO_3$ is considered to be environmental friendly since it is lead-free. There is another type of ferroelectric material that is much better in making sensors than $BaTiO_3$ which is plumbum titanate ($PbTiO_3$) but somehow $PbTiO_3$ is high in toxic which makes it not favorable to be used as it is harmful to environment. Nowadays,

$BaTiO_3$ have been widely studied because of their potential application in sensing and energy harvesting [26-28].

Many methods have been developed to synthesis $BaTiO_3$ films, such as pulse laser deposition (PLD) [29], metalorganic chemical vapor deposition (MOCVD) [30] and magnetron sputtering [31]. However, the aforementioned methods are difficult to scale up, since they require costly equipment and ultrahigh vacuum which limits the batch size and efficiency.

2. Experimental Details

In this research, textured of BaTiO3 films are synthesized through a two-step hydrothermal reaction. As shown in Figure 1, the TiO2 nanowires were first grown on Ti foil by a hydrothermal process. Ti foils were initially cleaned by sonication for 30 minutes in a mixed-solution of deionized water, acetone and 2-propanol with volume ratios of 1:1:1. Then the Ti foils were rinsed with methanol and deionized water. The Ti foils were then placed vertically inside an autoclave, which contained 10 mL of deionized water, 10 mL of hydrochloric acid and 1 mL of titanium isopropoxide. The autoclave was tightly closed and then placed in an oven at 160 ^{0}C for 4 hours. The resultant substrate with TiO2 nanowires on its surface was rinsed with deionized water and dried in ambient air. Subsequently, the TiO2 nanowires were converted to sol-gel method also can be applied to synthesis $BaTiO_3$. This method requires low cost and easy to handle. Nevertheless, this method has the bad sides where it requires high annealing temperatures which have high possibility to cause cracking when the thickness of the film is larger than a number of hundred nanometers [32].

A cost effective two-step hydrothermal method is introduced to grow dense $BaTiO_3$ films with controlled stoichiometry. The vertically aligned $BaTiO_3$ nanowires were synthesized by first preparing the vertically aligned TiO_2 nanowires as a template during the first step of hydrothermal reaction. The synthesized vertically aligned TiO_2 nanowires will then undergo a reaction with Ba^{2+} ions. The reaction will occur in the second step of hydrothermal reaction. The crystallinity, structure and morphology of the $BaTiO_3$ nanowires were characterized by X-ray diffraction (XRD) and scanning electron microscopy (SEM).

$BaTiO_3$ through a second hydrothermal reaction with an aqueous solution of $Ba(OH)_2.8H_2O$. After the reaction, the $BaTiO_3$ films were rinsed with deionized water and dried in ambient air. The samples were heated at 600 ^{0}C for 30 minutes to remove the hydroxyl ions which results due to the hydrothermal reaction conditions. Different reaction temperatures were applied to the samples during the first step of hydrothermal reaction to see the end result of the $BaTiO_3$ films.

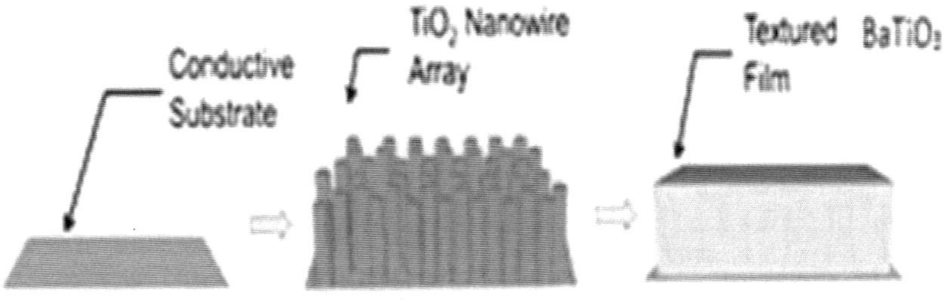

Figure 1. Schematic illustration to grow textured of $BaTiO_3$ films

3. Characterization

The crystalline structure of the synthesized TiO_2 and $BaTiO_3$ nanowires were analyzed by XRD using Rigaku Miniflex II Diffractometer with CuKα radiation (λ=1.5406), by scanning from angle of 20^0 to 80^0. The morphology and size of the grown TiO_2 and $BaTiO_3$ nanowires were characterized by SEM (JEOL) operated at 15 kV.

4. Results and Discussion

First step hydrothermal reaction: Formation of TiO_2 nanowires.

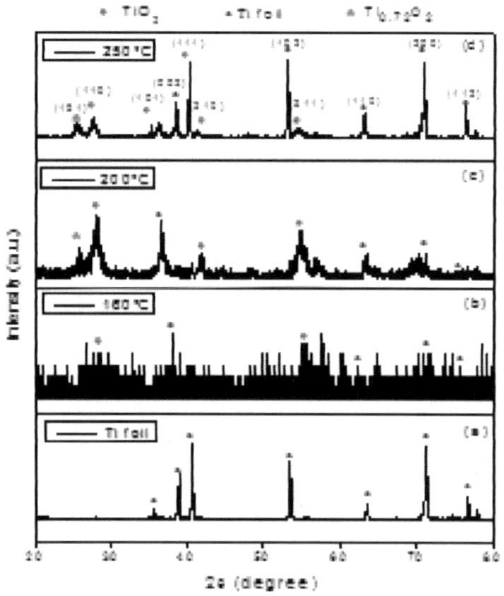

Figure 2. XRD patterns of (a) Ti foil and TiO_2 films at different reaction temperatures, (b) 160 ^{0}C, (c) 200 ^{0}C and (d) 250 ^{0}C.

Figure 2 shows the XRD pattern of first hydrothermal reaction of TiO_2 films prepared at different reaction temperatures namely 160 ^{0}C, 200 ^{0}C and 250 ^{0}C. Figure 2 (a) shows the XRD pattern of a bare TiO_2 film which is the substrate. When the sample was heated at 160 ^{0}C, the low peak intensity (as shown in Figure 2 (b)) is suggested that the TiO_2 nanowires were not fully formed and still in intermediate phase. The fingerprint of TiO_2 rutile pattern was clearly observed at 200 ^{0}C. With the increase of the first hydrothermal reaction temperature at 200 ^{0}C, the intermediate phase (Figure 2 (b)) turned into rutile structure (Figure 2 (c)). With the further increase of heated temperature up to 250 ^{0}C, the change of peak intensity of (111) at 2θ of 41.26 ^{0}, indicating the increased on crystalinity of TiO_2. Other rutile peaks are observed at 2θ of 27.37 ^{0}, 36.10 ^{0} and 56.36 ^{0} with *hkl* orientations of (110), (101) and (211).[36] However, the presence of minor $Ti_{0.72}O_2$ anatase (PDF No. 86-1157) was observed at 25.43 ^{0} with marginal peak intensity of (101). Based on the XRD patterns for Figure 2 (c) and (d), the form of major rutile phase and minor anatase phase were observed. Second step hydrothermal reaction: Formation of a layer of $BaTiO_3$ nanowires.

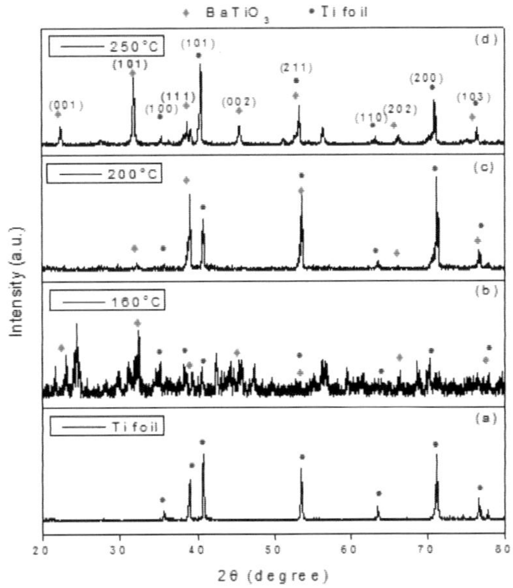

Figure 3. XRD patterns of (a) Ti foil and $BaTiO_3$ films at different reaction temperatures, (b) 160 ^{0}C, (c) 200 ^{0}C and (d) 250 ^{0}C.

Figure 3 shows the XRD pattern of $BaTiO_3$ films prepared from TiO_2 nanowires grown on TiO_2 film, which obtained via the first hydrothermal reaction at different reaction temperatures. The XRD pattern of the prepared $BaTiO_3$ samples as in Figure 3 could be readily indexed to tetragonal $BaTiO_3$ (PDF No. 75-2121) with a lattice parameter a = b = 3.9886 Å, c = 4.00060 Å. When the sample is heated with 160 ^{0}C, the peaks formed were scattered as shown in Figure 3 (b). The scattered peak indicates that $BaTiO_3$ films haven't form although the peaks started to show the little presence of $BaTiO_3$ phase. But the crystallinity of $BaTiO_3$ synthesised at 160 ^{0}C was low.

The fingerprint of $BaTiO_3$ rutile clearly observed at 200 ^{0}C. With the increase of the heated temperature, the intermediate phase (Figure 3 (b)) turned to rutile structure (Figure 3 (c)). With the further increase of heated temperature up to 250 ^{0}C, some of the peak intensity were enhanced such as at (001), (101), (111), and (202). Whereas others disappeared which are for peak intensity of (101) and (211). At 250 ^{0}C, the change of peak intensity of (101) $BaTiO_3$ at 2θ of 31.62 ^{0} was clearly observed, which affirmed the formation of high crystalline $BaTiO_3$.[37] Instead of the peak showed $BaTiO_3$ and Ti foil, presence of BaO was also determined.

Morphology of TiO_2 films (First step hydrothermal reaction at 200 ^{0}C).

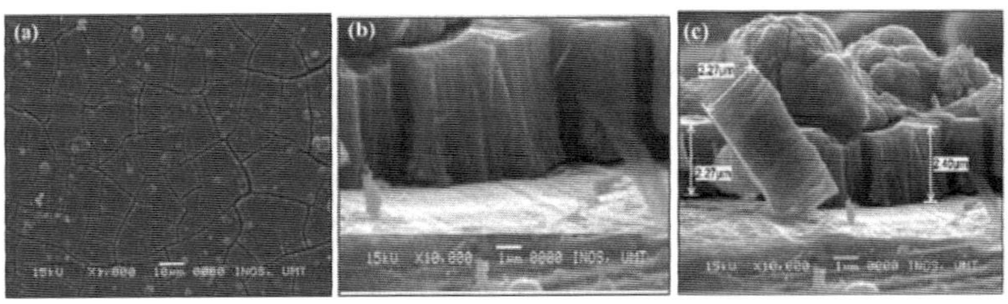

Figure 4. (a) TiO_2 surface morphology (b) TiO_2 cross sectional area (c) Thickness of TiO_2

For SEM analysis of TiO_2 films, only one sample was measured which the sample is for optimum temperature. The synthesis of the desired $BaTiO_3$ films begins with the growth of densely packed TiO_2 nanowires on the surface of Ti foil as shown in Figure 4 through the first hydrothermal reaction step. When observed in cross sectional as in Figure 4 (b), the top of the TiO_2 films has more free space and less dense than the base of TiO_2 nanowires. This indicated that there were sufficient space for the infusion of Ba^{2+} ions from the precursor used in the second hydrothermal reaction to react with TiO_2 nanowires. The grain and length of TiO_2 nanowires can be controlled and modified by differ reaction heating time in the first hydrothermal reaction as shown in Figure 4 (a). The first step hydrothermal reaction causes the vertically alligned TiO_2 nanowires grow on the Ti foil surface at 200 ^{0}C.

Morphology of $BaTiO_3$ films (Second step hydrothermal reaction at 160 ^{0}C)

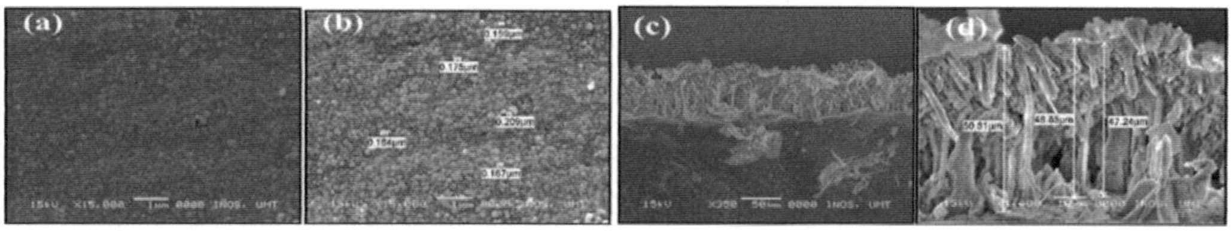

Figure 5. (a) $BaTiO_3$ surface morphology (b) Width of $BaTiO_3$ (c) $BaTiO_3$ cross sectional area (d) Thickness of $BaTiO_3$

Figure 4 (a) showed the resultant layer of vertically alligned TiO_2 nanowires was obtained after the first step of hydrothermal reaction. We believed that 200 ^{0}C is the optimum temperature for the growth of TiO_2 nanowires on the Ti foil surface. It is because, the TiO_2 films prepared a 200 ^{0}C showed the structure of nanowires as compared to other temperatures which are 160 ^{0}C and 250 ^{0}C. From the Figure 4 (a), it is obviously observed that the surface morphology of the TiO_2 nanowires is flaky and forming grains. The thickness of the TiO_2 nanowires in the range of 2.27 μm to 2.40 μm was measured as depicted in Figure 4 (c). Observation of the surface morphology of $BaTiO_3$ sample heated at 160 ^{0}C showed the elongation of the vertically alligned $BaTiO_3$ nanowires. Figure 5 (a) also showed the surface morphology of $BaTiO_3$ nanowires that synthesised by hydrothermal reaction with 4 hours soaking time at heating temperature of 160 ^{0}C. In the second hydrothermal reaction, the growth of $BaTiO_3$ structure formed nanorod rather than nanowire as shown in Figure 5 (c) and (d). The transformation of nanorod to nanowire occurs due to low temperature which might causes partial reaction. The cross sectional area and the results proved that the heights of $BaTiO_3$ nanowires were not uniform. The height of the $BaTiO_3$ nanowires were in the range of 47.240 μm to 50.810 μm. The mean of the height is 48.970 μm. The width of the $BaTiO_3$ nanowires was measured with a mean value of 0.179 μm in the range of 0.164 μm to 0.209 μm.

Morphology of $BaTiO_3$ films (Second step hydrothermal reaction at 200 ^{0}C).

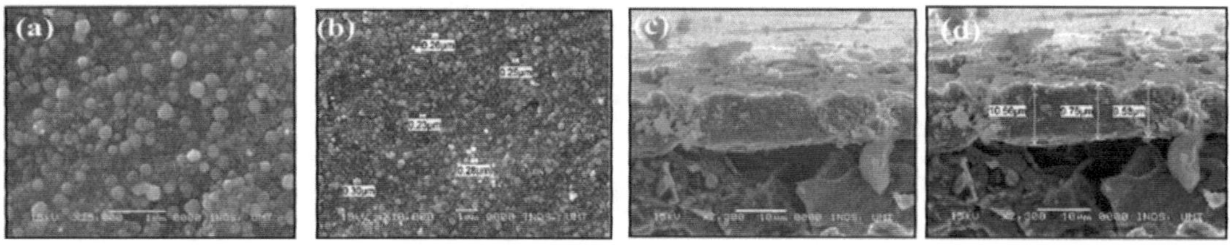

Figure 6. (a) $BaTiO_3$ surface morphology (b) Width of $BaTiO_3$ (c) $BaTiO_3$ cross sectional area (d) Thickness of $BaTiO_3$

Observation of the surface morphology of $BaTiO_3$ that heated at 200 ^0C showed that the height of the vertically alligned $BaTiO_3$ nanowires are slightly to be uniformed. With the increase of heating temperature up to 200 ^0C in the second hydrothermal reaction, the growth of $BaTiO_3$ structure is fully forming nanowires with spacing rather than forming dense $BaTiO_3$ nanowires. The width of the $BaTiO_3$ nanowires was measured with the mean value of 0.264 μm in the range of 0.230 μm to 0.30 μm as shown in Figure 6 (b). The cross sectional area as in Figure 6 (c) proved that the height of the vertically alligned $BaTiO_3$ nanowires were not uniform and in the range of 9.580 μm to 10.560 μm. The mean value is 9.9630 μm.

Morphology of $BaTiO_3$ films (Second step hydrothermal reaction at 250 ^0C).

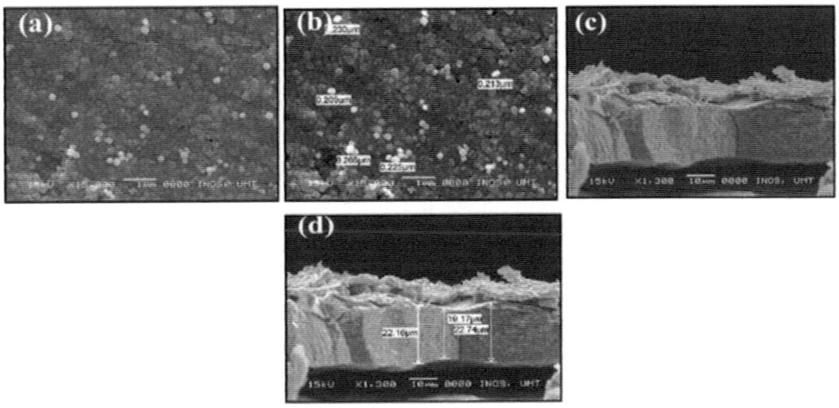

Figure 7. (a) $BaTiO_3$ surface morphology (b) Width of $BaTiO_3$ (c) $BaTiO_3$ cross sectional area (d) Thickness of $BaTiO_3$

Observation of the surface morphology of $BaTiO_3$ sample that heated at 250 ^0C showed that the height of the vertically alligned $BaTiO_3$ nanowires to be more uniform compared to 160 ^0C and 200 ^0C. The width of the $BaTiO_3$ nanowires was measured with the mean value of 0.229 μm in the range of 0.209 μm to 0.266 μm. The cross sectional area as in Figure 7 (c) indicated that the height of the vertically alligned $BaTiO_3$ nanowires were obviously not uniform and they are very dense. The height of the $BaTiO_3$ nanowires were in the range of 19.170 μm to 22.740 μm with the mean value of 21.357 μm.

5. Conclusion

Single phase BaTiO3 films were successfully fabricated by using a two-step hydrothermal reaction. The Ti foil is very suitable to grow the BaTiO3 films on it through the use of vertically aligned TiO2 nanowires as the template. The crystalinity of the samples were seen throughout the XRD characterization and it was found out that the scattered peak at heating reaction of 160 ^0C in the first and second hydrothermal reaction happened because of the partial reaction occurred at that temperature. For the SEM analysis, the image of nanowires were seen to form at the heating temperature of 200 ^0C. This has made the temperature as the optimum temperature to grow the desired nanowires. Besides, this hydrothermal method offers a low cost method to produce thin films for microelectronics.

Acknowledgement

We would like to thank Ibnu Sina Institute for Scientific and Industrial Research (ISI-SIR), Universiti Teknologi Malaysia (UTM) for providing research facilities. This research work has supported by PDRU (03E22) and Tier 1 (16H42) grants.

References

[1] Tressler J F, Alkoy S and Newnham R E J 1998 *Electroceram.* **2**(4) 257-272
[2] Mossi K M, Selby G V and Bryant R G 1998 *Mater. Lett.* **35** 39-49
[3] Uchino K 1996 Piezolectric actuators and ultrasonic motors. *Springer: New York, 1*
[4] Damjanovic D and Newnham R J 1992 *Intell. Ater. Syst. Struct.* **3** 190-208
[5] Uchino K 2000 In Ferroelectric devices *CRC: Boca Raton, FL* **16**
[6] Muralt P J, Micromech and Microeng 2000 **10** 136-146
[7] Cook-Chennault K, Thambi N and Sastry A 2008 *Smart Mater. Struct.* **17**(4) 043001
[8] Priya S J 2007 *Electroceram.* **19**(1) 167-184
[9] Cha S, Kim S M, Kim H, Ku J, Sohn J L, Park Y J, Song B G, Jung M H, Lee E K, Choi B L, Park J J, Wang Z L, Kim J M and Kim K 2011 *Nano Lett.* **11** 5142-5147
[10] Wu J, Wang J, Xiao D and Zhu J 2011 *ACS Appl. Mater. Interfaces* **3** 3261-3263
[11] Zhang S, Han M, Zhang J, Li Y, Hu Z and Chu J 2013 *ACS Appl. Mater. Interfaces* **5** 3191-3198
[12] Liang W, Ji Y, Nan T, Huang J, Bi Z, Zeng H, Du H, Chen C, Jia Q and Lin Y 2012 *ACS Appl.Mater. Interfaces* **4** 2199-2203
[13] Wu J, Wang J, Xiao D and Zhu J 2011 *Electrochem. Solid-State Lett.* **14** G57-G59
[14] Tang H, Lin Y, Andrews C and Sodano H A 2011 *Nanotechnology* **22** 015702
[15] Datta A, Mukherjee D, Hordagoda M, Witanachchi S, Mukherjee P, Kashid R V, More M A, Joag D S and Chavan P G 2013 *ACS Appl. Mater. Interfaces* **5** 6261-6267
[16] Liu M, Li X, Imrane H, Chen Y, Goodrich T, Cai Z, Ziemer K S, Huang J Y and Sun N X 2007 *Appl. Phys. Lett.* **90** 152501
[17] Son J Y, Lee J, Song S, Shin Y, Jang H M 2013 *ACS Nano* **7** 5522-5529
[18] Narayanan T N, Mandal B P, Tyagi A K, Kumarasiri A, Zhan X, Hahm M G, Anantharaman M R, Lawes G and Ajayan P M 2012 *Nano Lett.* **12** 3025-3030
[19] Zhu Y, Zhang L, Natsuki T, Fu Y and Ni Q 2012 *ACS Appl. Mater. Interfaces* **4** 2101-2106
[20] Szwarcman D, Vestler D and Markovich G 2011 *ACS Nano* **5** 507-515
[21] P M Rorvik, T Grande and M A Einarsrud 2011 One-dimensional nanostructures of ferroelectric perovskites. *Adv. Mater.* **23** 4007-4034
[22] N Bao, L Shen, G Srinivasan, K Yanagisawa and A Gupta 2008 Shape-controlled monocrystalline ferroelectric barium titanate nanostructures: from nanotubes and nanowires to ordered nanostructures. *J. Phys. Chem. C.* **112** 8634-8642
[23] L Torrente-Murciano, A A Lapkin and D Chadwick 2010 Synthesis of high aspect ratio titanate nanotubes. *J. Mater. Chem.* **20** 6484-6489
[24] Zhang X, Zhao X, Lai C, Wang J, Tang X and Dai J *Appl. Phys. Lett. 24,* **85**, 4190-4193
[25] Lin Y, Liu Y and Sodano H A 2009 *Appl. Phys. Lett.* **95** 122901
[26] Limmer S J, Seraji S, Wu Y, Chou T P, Nguyen C and Cao G Z 2002 *Adv. Func. Mater.* **12** 59-64
[27] Huang K, Huang T and Hsieh W 2009 *Inorg. Chem.* **48** 9180-9184
[28] Teresa Buscaglia M, Harnagea C, Dapiaggi M, Buscaglia V, Pignolet A and Nanni P 2009 *Chem. Mater.* **21** 5058-5065
[29] J A Bellotti, W Chang, S B Qadri, S W Kirchoefer and J M Pond 2006 *Appl. Phys. Lett.* **88** 012902

[30] Y S Min, Y J Cho, D Kim, J H Lee, B M Kim, S K Lim, I M Lee and W I Lee 2001 *Chem. Vap. Deposition* **7** 146-1449
[31] T Horikawa, N Mikami, T Makita, J Tanimura, M Kataoka, K Sato and M Nunoshita, Jpn J 1993 *Appl. Phys.* **32** 4126
[32] A Atkinson and R Guppy J 1991 *Mater. Sci.* **26** 3869-3873
[33] J G Lisoni, C H Lei, T Hoffman and V M Fuenzalida 2002 *Surf. Sci.* **515** 431-440
[34] Z Wu, N Kumagai and M Yoshimura 2000 *Chem. Mater.* **12** 3356-3361
[35] D Chen, H Zhang, R Chen, X Deng, J Li, G Zhang and L Wang 2002 *Phys. Status Solidi* **209** 714-717
[36] Venkatachalam S, Hayashi H, Ebina T and Nanjo H 2013 Preparation and Characterization of Nanostructured TiO2 Thin Films by Hydrothermal and Anodization Methods. In S. L. Pyshkin & J.M. Ballato (Eds.), *Optoelectronics Advanced Materials and Devices* (Ch.05). Rijeka: InTech
[37] Deng Z, Dai Y, Chen W, Pei X and Liao J 2010 Synthesis and characterization of bowl-like single-crystalline BaTiO3 nanoparticles. *Nanoscale research letters* **5**(7) 1217

IOP Conf. Series: Journal of Physics: Conf. Series **1027** (2018) 012015 doi:10.1088/1742-6596/1027/1/012015

Calcium to phosphate ratio measurements in calcium phosphates using LIBS

U Tariq*[1], Z Haider[1], K Chaudhary[1,2], R Hussain[3] and J Ali[1,2]

[1]Laser Center, Ibnu Sina Institute for Scientific and Industrial Research (ISI-SIR), Universiti Teknologi Malaysia, 81310 UTM Johor Bahru, Johor, Malaysia
[2]Physics Department, Faculty of Science, Universiti Teknologi Malaysia, 81310 UTM Johor Bahru, Johor, Malaysia
[3]Department of Physics, COMSATS Institute of Information Technology, Islamabad, Pakistan

Corresponding author: usmaan.t@gmail.com

Abstract. In this study, we report the measurement of Ca/P ratio of hydroxyapatite using LIBS. Hydroxyapatite was prepared by wet chemical precipitation method with different stoichiometric ratios of Ca and P (1.5, 1.6, 1.67, 1.70, 1.80). Samples were then pelletized using KBr as a binder under a hydraulic press. Samples were irradiated with a laser beam and corresponding spectra were recorded. Spectra were further analyzed to calculate the Ca/P ratio which was then compared with the EDX for validation. LIBS results showed promise with a deviation of < 5% with the Stoichiometric ratios and EDX results.

1. Introduction

Natural bone is a composite of mineral and organic phase (70% mineral and 30% organic). There is immense importance of hydroxyapatite in biomedical applications as 70% mineral phase of natural bone contains 95% hydroxyapatite [1]. Biomaterials can be prepared synthetically to function appropriately in a bio-environment. After the invention of first generation of materials during 1960–1970 for use inside a human body, synthetic biomaterials became a subject of interest [2]. Calcium phosphate ceramics are usually described by their calcium/phosphorus atomic ratio (Ca/P). Among these ceramics, hydroxyapatite ($Ca_{10}(PO_4)_6(OH)_2$, HA, Ca/P = 1.667) and β-tricalcium phosphate (β-$Ca_3(PO_4)_2$, β-TCP, Ca/P = 1.5) have been extensively studied, and their various biological behaviours are well-known [3].

The properties of calcium phosphates of biological interest depend strongly on their calcium/phosphorus atomic ratios (Ca/P). Therefore, the precision for the determination of this ratio is crucial [3].Calcium to phosphate ratio plays a very important role in the mechanical properties of the biomaterials. In a previous study [3] it is shown that a small variation in the Ca/P ratio of calcium deficient HA from 1.650 to 1.667 (i.e relative deviation of Ca/P ratio <1%) resulted in substantial changes in the mechanical characteristics in densified BCP ceramics. Various techniques XRD[2] , EDX [4,5], XRF [6,7], ICP [8] have been employed for elemental analysis and to find the Ca/P ratio of hydroxyapatite. All these techniques require special sample preparation and most of them require vacuum conditions and are time consuming. LIBS gives the advantage of minimal or no sample preparation along with being a rapid measurement technique [9].

LIBS is a powerful and versatile analytical technique for elemental analysis [10]. It is an optical emission spectroscopy of laser induced plasma which is a source of ions and electrons [11]. In LIBS, laser pulse interacts with a target generating plasma [12].For chemical analysis LIBS is a highly potential diagnostic technique [13].

In this study, we have used LIBS to determine the Ca/P ratio in synthetically prepared hydroxyapatite with Ca/P ratios ranging from 1.5-1.8. Suitable peaks of Ca and P are selected for the reliable determination of Ca/P ratio.

2. Experimental

2.1. Sample Preparation fpr LIBS

Calcium nitrate tetrahydrate (Ca(NO3)2.4H2O, QReC) and di ammonium hydrogen phosphate ((NH4)2HPO4, QReC) were used as precursors for Calcium and Phosphate elements respectively. Corresponding molar solution of calcium nitrate tetrahydrate and diammonium hydrogen phosphate were prepared in 100 ml DDW. Solution of di ammonium hydrogen phosphate was added dropwise in calcium nitrate tetrahydrate solution with continuous stirring. The reaction mixture was then transferred to the 800 W household microwave (SHARP, model R-218LS) to reflux for 30 min [5]. Resulting precipitates were filtered, washed with distilled water, dried in the oven at 80 °C for 17 h and calcined in muffle furnace at 1000 °C for 2 h. Five samples (A, B, C, D & E) of HA were prepared with Ca/P ratio 1.5, 1.6, 1.67, 1.7 and 1.8 respectively.

After completing the calcination process, the mixture of synthetic HA and binder (KBr) of known quantities was pelletized using hydraulic press by pressing by applying 10 tons force to compress the sample as shown in Figure 1. The pellets were prepared in identical mass and dimensions.

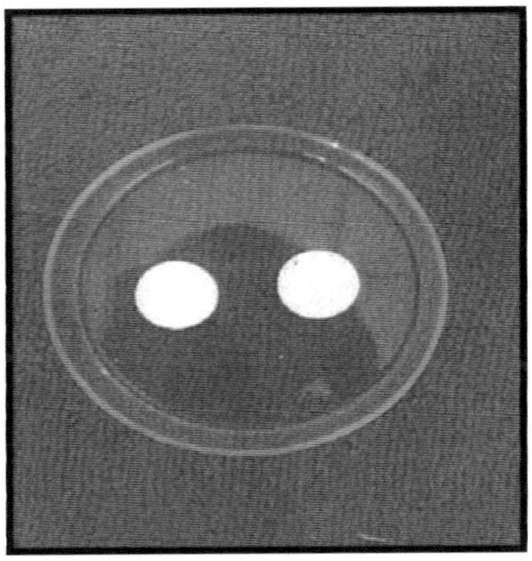

Figure 1. Image of one of synthetic HA samples in pellet form

2.2. LIBS Setup and Procedure

The LIBS experimental setup shown in Figure 2 was used for LIBS analysis of samples. Nd:YAG (1064 nm, 10 ns, 10 Hz) laser pulses of 320 mJ of energy were focused onto the target sample, a fiber optic cable was used to collect plasma radiations and transfer to Ocean Optics MayaPro spectrometer equipped with CCD detector. Laser pulse energy was constantly examined by deflecting 6% of the beam to a calibrated energy meter through quartz beam reflector.

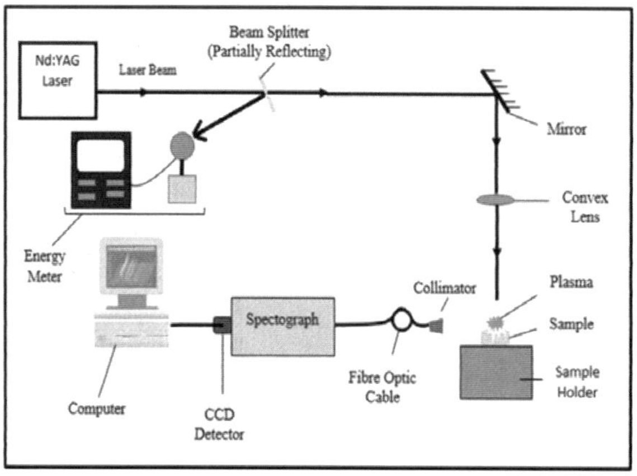

Figure 2. Schematic diagram setup for Nd:YAG laser pulse using LIBS

Each sample was exposed at 10 distinct spots on the surface to 10 laser shots at each spot. For each laser shot an emission spectrum from the plasma was recorded on the computer. Ca/P ratios were calculated using peak intensities of Ca and P emission lines. Results were compared with EDX results, which were considered as reference, to calculate an approximate error in the results obtained from LIBS.

3. Results and Discussion

3.1. EDX

Energy Dispersive X-ray (EDX) spectroscopy was used for the elemental analysis of prepared samples. Figure 3 shows the EDX graph of the sample D.

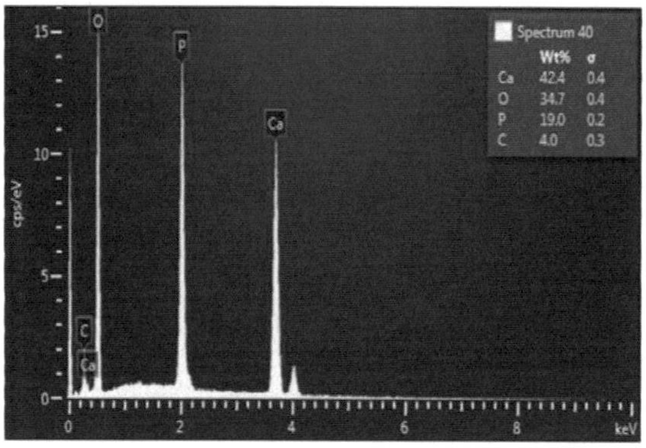

Figure 3. EDX graph of sample D

Four spots on each sample were used for the EDX analysis. Ca/P ratio of each sample was calculated as an average of these four spots. EDX analysis showed that the Ca/P ratio of the samples investigated in this study were close to their stoichiometric ratio. This confirmed that the samples are appropriately prepared in desired Ca/P ratios. Table 1 shows Ca/P ratios in the samples obtained from EDX analysis.

Table 1. Results of calculation of Ca/P ratio based on EDX analysis.

Sample	Stoichiometric Ca/P ratio	Average of weight percentage of Ca (%)	Average of weight percentage of P (%)	EDX Ca/P ratio
A	1.50	33.30	17.00	1.52
B	1.60	37.73	18.10	1.62
C	1.67	41.63	18.93	1.70
D	1.70	42.50	19.05	1.73
E	1.80	42.93	18.20	1.83

3.2. LIBS Analysis

LIBS spectrum of the sample D is shown in figure 4. Theoretically, intensity of spectral lines in emission spectra is indicative of elemental concentration. However, different spectral lines of same element do not respond equally to a similar change in the element's concentration. Ca/P ratio was therefore calculated by intensity ratio of all possible combinations of Ca and P lines identified in the LIBS spectra. Lines of Ca, P, K and Br were identified in the spectrum. The identification of the spectral lines is performed using NIST atomic spectra database [http://physics.nist.gov/PhysRefData/ASD/lines_form.html].

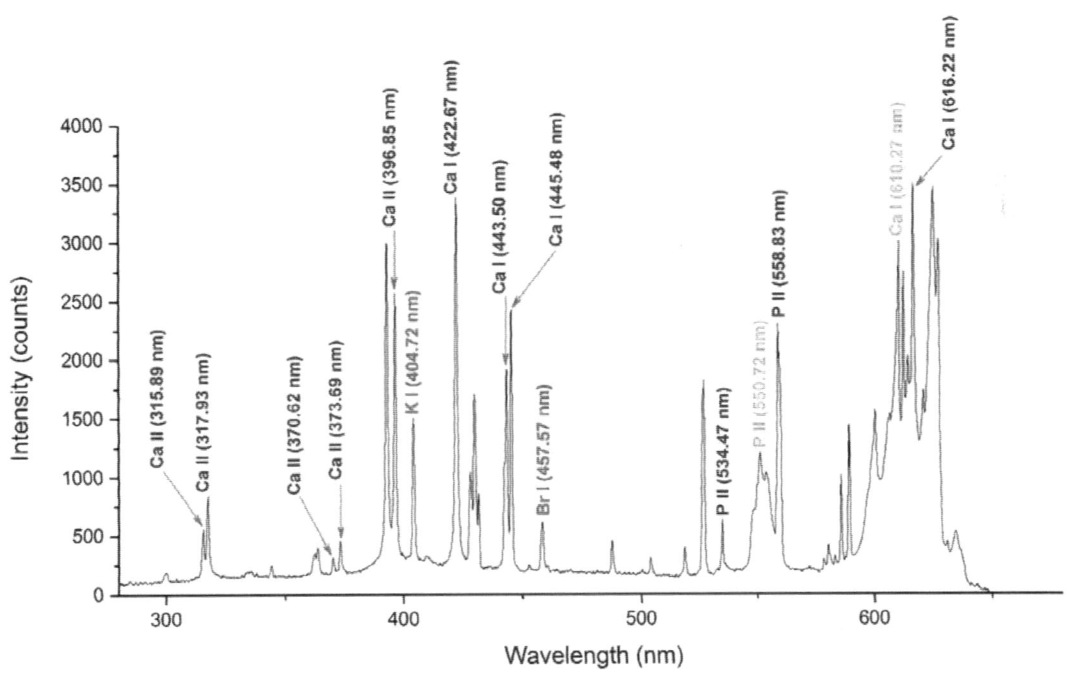

Figure 4. LIBS spectra of sample D.

Of all the combinations of Ca and P lines used for calculating Ca/P ratio, Ca I 610.27 nm and P I 550.72 nm gave closed values to the stoichiometric values (also verified by EDX). Same pair of lines

IOP Conf. Series: Journal of Physics: Conf. Series **1027** (2018) 012015 doi:10.1088/1742-6596/1027/1/012015

is used to determine Ca/P ratio in all five samples. Emission lines are authenticated from NIST database [14] and the published literature [15]. The calculation of Ca/P ratio based on LIBS results for all samples are given in Table 2.

Table 2. Results calculation of Ca/P ratio based on LIBS results.

Sample	Stoichiometric Ca/P ratio	Average intensity of Ca line	Average intensity of P line	LIBS Ca/P ratio
A	1.50	1815.02	1184.68	1.53
B	1.60	4951.48	3080.92	1.61
C	1.67	3371.64	1984.78	1.70
D	1.70	2996.38	1744.17	1.72
E	1.80	3963.22	2179.09	1.82

EDX analysis has been used as the reference to LIBS analysis since EDX technique has already commercialized technique for such analysis. Table 3 shows comparison of Ca/P ratio for all five samples that analysed using LIBS and EDX techniques.

Table 3. Comparison of Ca/P ratio for all five samples that are analyzed using LIBS and EDX techniques

Sample	Stoichiometric Ca/P ratio	LIBS Ca/P ratio	EDX Ca/P ratio
A	1.50	1.53	1.52
B	1.60	1.61	1.62
C	1.67	1.70	1.70
D	1.70	1.72	1.73
E	1.80	1.82	1.83

Results from the table 3 show that, the estimation of LIBS results are comparable with EDX results. Figure 5 shows the graph of Ca/P ratio calculated by EDX and LIBS in comparison with stoichiometric values.

Samples A, B, C, D and E are synthetic HA samples with Ca/P ratio of 1.50, 1.60, 1.67, 1.70 and 1.80 respectively. From graph, LIBS Ca/P ratio (red line) is close to EDX Ca/P ratio (blue line). The difference between LIBS and EDX Ca/P ratio is only ±0.01. These shows that Ca/P ratio of synthetic HA can be calculated using LIBS technique and the results could be nearly as accurate as those obtained from EDX.

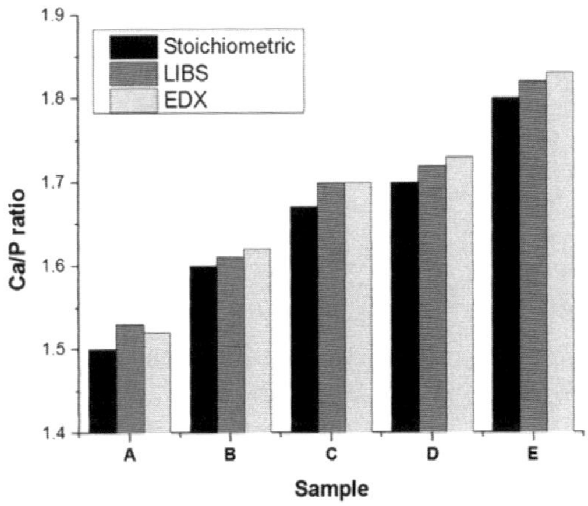

Figure 5. Graph of Ca/P ratio of prepared samples

4. Conclusion

In this study LIBS has been successfully utilized as a fast and reliable technique to find the Ca/P ratio of synthetic hydroxyapatite. The comparison of results of LIBS and EDX have demonstrated that it is a potential candidate to be used as an alternative technique for the elemental qualitative as well as quantitative study of calcium phosphates. To verify the authenticity of this further studies are required to replicate these results.

Acknowledgement

We would like to thank Ibnu Sina Institute for Scientific and Industrial Research (ISI-SIR), Universiti Teknologi Malaysia (UTM) for providing research facilities. This research work has supported by PDRU (03E22) and FRGS (4F891) Grants.

References

[1] Alvarez K and Nakajima H 2009 Metallic Scaffolds for Bone Regeneration. *Materials* **2**(3) 790–832

[2] Hench L L 2005 Bioceramics. *Journal of the American Ceramic Society* **81**(7) 1705–1728

[3] Raynaud S, Champion E, Bernache-Assollant D and Laval J 2001 Determination of Calcium/Phosphorus Atomic Ratio of Calcium Phosphate Apatites Using X-ray Diffractometry. *Journal of the American Ceramic Society* **84**(2) 359–366

[4] Ofudje E A, Rajendran A, Adeogun A I, Idowu M A, Kareem S O and Pattanayak D K 2017 Synthesis of Organic Derived Hydroxyapatite Scaffold from Pig Bone Waste for Tissue Engineering Applications. *Advanced Powder Technology*

[5] Taha A, Akram M, Jawad Z, Alshemary A Z and Hussain R 2017 Strontium Doped Injectable Bone Cement for Potential Drug Delivery Applications. *Materials Science and Engineering C* **80** 93–101

[6] Barakat N A M, Khil M S, Omran A M, Sheikh F A and Kim H Y 2009 Extraction of Pure Natural Hydroxyapatite from the Bovine Bones Bio Waste by Three Different Methods. *Journal of Materials Processing Technology* **209**(7) 3408–3415

[7] Alshemary A Z, Akram M, Goh Y-F, Tariq U, Butt F K, Abdolahi A and Hussain R 2015 Synthesis, Characterization, in Vitro Bioactivity and Antimicrobial Activity of Magnesium and Nickel Doped Silicate Hydroxyapatite. *Ceramics International* **41**(9) 11886–11898

[8] Saleh A T, Ling L S and Hussain R 2016 Injectable Magnesium-Doped Brushite Cement for Controlled Drug Release Application. *Journal of Materials Science* **51**(16) 7427–7439

[9] Tariq U, Haider Z, Hussain R, Tufail K and Ali J 2017 LIBS Analysis of Hydroxyapatite Extracted from Bovine Bone for Ca/P Ratio Measurements. In *International Conference on Plasma Science and Applications (ICPSA 2016)*; *American Institute of Physics* **1824** 30027

[10] Chaudhary K, Rizvi S Z H, Ali J 2016 Laser-Induced Plasma and Its Applications. In *Plasma Science and Technology - Progress in Physical States and Chemical Reactions*. InTech.259–291

[11] Khaleeq-ur-Rahman M, Bhatti K A, Rafique M S, Latif A and Chaudhary K T 2009 Magnetic Field Effect on Electron Emission from Plasma. *Vacuum* **83**(6) 936–941

[12] Arab M, Bidin N, Rizvi Z H, Safie S and Alsaedi M A 2014 Comparison Study of Two Commercial Spectrometers for Heavy Metal Analysis of Laser Induced Breakdown Spectroscopy (LIBS). *Photonic Sensors* **4**(1) 63–69

[13] Chaudhary K, Rosalan S, Aziz M S, Bohadoran M, Ali J, Yupapin P P and Bidin N 2015 Saktioto. Laser-Induced Graphite Plasma Kinetic Spectroscopy under Different Ambient Pressures. *Chinese Physics Letters* **32**(4) 43201

[14] NIST database https://www.nist.gov/pml/atomic-spectra-database

[15] Haider Z, Ali J, Arab M, Munajat Y, Roslan S, Kamarulzman R and Bidin N 2016 Plasma Diagnostics and Determination of Lead in Soil and Phaleria Macrocarpa Leaves by Ungated Laser Induced Breakdown Spectroscopy. *Analytical Letters* **49**(6) 808–817

IOP Conf. Series: Journal of Physics: Conf. Series **1027** (2018) 012016 doi:10.1088/1742-6596/1027/1/012016

The application of sensitizers from red frangipani flowers and star gooseberry leaves in dye-sensitized solar cells

Wan Almaz Dhafina[1], Hasiah Salleh[2]*, Muhamad Zalani Daud[1] and Nora'aini Ali[1]

[1]School of Ocean Engineering, Universiti Malaysia Terenggganu, 21030 Kuala Nerus, Terengganu, Malaysia
[2]Centre for Fundamental and Liberal Education, Universiti Malaysia Terengganu, 21030 Kuala Nerus, Terengganu, Malaysia

Corresponding author: hasiah@umt.edu.my

Abstract. Nowadays natural based dyes for dye-sensitized solar cells (DSSCs) have been in research field attention due to its advantages over other type of dyes such as low-cost, low-toxicity, completely biodegradable and abundance of resources. Natural dyes can be produced via the simple extraction method of pigments from plant parts such as flower, fruits, leaves, tuber etc. In this feature article, the natural dyes which composed of anthocyanin pigment from red frangipani flowers and chlorophyll from star gooseberry leaves were applied in zinc oxide, (ZnO) based-DSSC. The ZnO photoanode of the DSSCs sample were sensitized in each dye with different duration. It was observed that DSSCs which has chlorophyll pigment as dye had better performance with power conversion efficiency (PCE) of 0.007%.

1. Introduction

DSSCs is among third generation solar cell that works with the same principal with photochemical cell, but depends on the photon from sunlight to make chemical reaction and generates energy. It was pioneered by Gratzel's group which its architecture consists of nanoparticle of titania, ruthenium based dye and iodide species electrolyte [1]. For the record, N3 dye (one of ruthenium dye) shows constant high PCE compared to the other kind of dyes until now. However, the ruthenium dye is high-cost due to the lack of ruthenium resource, complex synthetic route and purification requirement [2]. Due to this problem, researches searched new ways to substitute ruthenium based dye and lead to the finding of alternative dyes which are organic dye and natural dye. These two dyes are low-cost and rather environmental friendly. Unfortunately, the synthetic route of organic dye is very complicated, and the yield is very low [3]. In other hand, natural dye is easy to be produced via simple extraction of plant pigments [4] using conventional solvents. Mostly, metal oxide of choice used in DSSC is titania but several studies have attempted to use zinc oxide (ZnO) due to low-cost [5], higher electron mobility [6] and higher conduction band energy level that prompt the better electron injection. It has been reported that ZnO is the first material used in experimental design to proof the irreversible electron injection from organic molecules into the conduction band of metal oxide [7]. Basically, DSSC architecture composed of photoanode and cathode that sandwiched together with electrolyte fillings. Several requirements need to be met regarding the energy level of every materials in DSSC: 1) The highest occupied molecular orbital (HOMO) of chosen dye must be lower than redox level, 2) the lowest unoccupied molecular orbital (LUMO) of chosen dye should be sufficiently higher than

IOP Conf. Series: Journal of Physics: Conf. Series **1027** (2018) 012016 doi:10.1088/1742-6596/1027/1/012016

conduction band of metal oxide of choice and 3) catalyst metal of choice in electrode must not chemically reactive with electrolyte and dye.

In this feature article, we discussed about sensitization of ZnO nanoparticles and natural dyes from red frangipani flowers, noted as red frangipani dye (RFD) and star gooseberry leaves, noted as star gooseberry dye (SBD). To the best of our knowledge, there is no research has been done to investigate the potential of combination between ZnO and RFD and there are no reports has been made regarding SBD.

2. Experimental

2.1. Fabrication of ZnO thin film

The indium tin oxide (ITO) glass substrate with dimension of 2cm x 2cm was use as a base to fabricate active materials in photoanode. The ITO glasses were cleaned ultrasonically in distilled water, acetone (R&M Chemicals) and ethanol ethanol (95%, HmbG Chemicals) for 5 min, respectively and dried with hairdryer [8]. The ZnO nanopowder, Sigma-Aldrich (0.025M) in methanol, (R&M Chemicals) was spin-coated with spin-coater, MODEL WS-400B-6NPP/LITE on ITO glass with ten repetitions at velocity of 2000rpm for 20s [9]. After that the ZnO thin film was dried with hair dryer and kept in dry box.

2.2. FRD and SBD preparation

Both RFD and SBD were prepared via simple extraction method using ethanol with the ratio between flowers and leaves over ethanol was 1/10. The flowers of red frangipani and star gooseberry leaves (Figure 1) were cut into small pieces, washed with distilled water several times and blew dried with hairdryer. After that the flowers and leaves were immersed respectively in ethanol and heated at 40°C for 2 hours and stirred at 350 rpm. After that the solution were filtered and the crude dye was used as it is (Figure 1(c)). The structure of anthocyanin pigment from RF flowers and chlorophyll from SB leaves are showed in Figure 1(d) and 1(e) respectively [10].

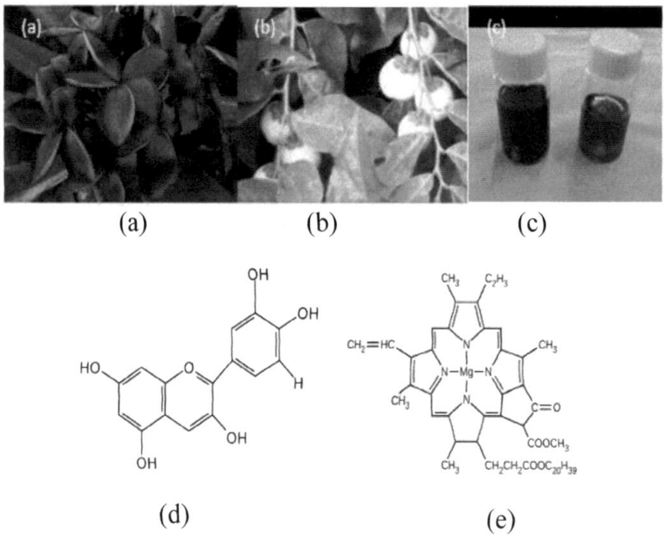

Figure 1. (a) red frangipani flowers, (b) star gooseberry leaves, (c) extracted dyes solution (the green one is SBD while the red one is RFD), (d) Chemical structure of anthocyanin and (e) chemical structure of chlorophyll [11].

2.3. Dye sensitizing on ZnO thin film
The ZnO thin films were sensitized by dipping them in RFD and SBD respectively in dark environment. The sensitization process was varied from 2h-8h with 2h interval. After that, sensitized thin film was taken out, rinsed with ethanol several times and dried with hairdryer.

2.4. Penetration of cathode and electrolyte
To prepare cathode, clean ITO glass was sputtered with aurum with thickness of 0.001nm. Electrolyte solution was prepared by mixing 0.5M of Lithium Iodide (LiI), Sigma-Aldrich, 99.9% trace metals basis and 0.05 Iodine (I2), Fluka Analytical in acetonitrile, (Merck KGaA) and kept in the dark environment.

2.5. Assemble of DSSC
Both electrodes (cathode and photoanode) were sandwiched together (Figure 2) with a piece of parafilm that have punched hole with diameter of 6mm inserted between electrodes. The electrolyte was dropped through parafilm hole on photoanode before covered with cathode. After that both electrodes were clipped together with mask with punched hole of 6mm diameter.

Figure 2. Sandwiched cathode and photoanode of DSSC. Where cathode was ITO/Au and anode was ITO/ZnO/dye+ electrolyte.

2.6. Characterization
RFD and SBD were characterized in several aspects. UV-Vis spectrometer (Model Lambda 25 with UV Winlab V2.85 software, Perkim Elmer) was employed to characterize optical properties of RFD and SBD while, the functional group of RFD and SBD were investigated via Fourier transform infrared (FTIR). The HOMO and LUMO level of RFD and SBD were investigated through cyclic–voltammetry (C-V) from METROHM AUTOLAB and lastly the power conversion efficiency (PCE) of DSSCs were studied by employing KEITHLEY 4200-SCS SEMICONDUCTOR CHARACTERIZATION SYSTEM.

3. Results and discussion

3.1. UV-Vis absorption and determination of Functional Groups via Fourier Transforms Infra-Red of RFD and SBD
Absorption spectra of RFD and SBD are shown in Figure 3(a). Max absorption peak of RFD was at 542nm which proved the existent of anthocyanin pigment content [12-13] while absorption spectra pattern of SBD was matched with results reported by [14] which proved it was chlorophyll pigments that consist of chlorophyll a and chlorophyll b.

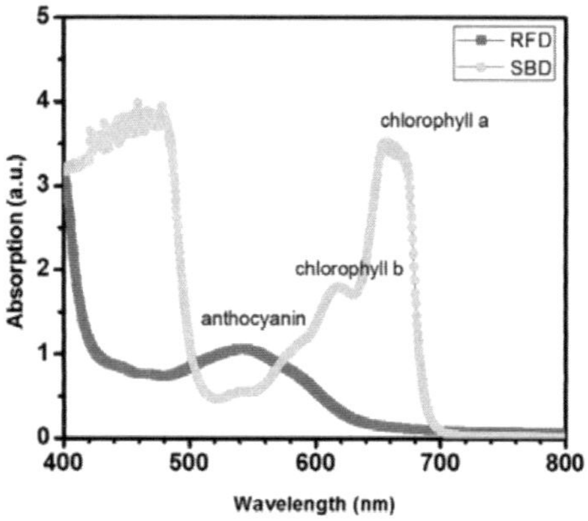

(a)

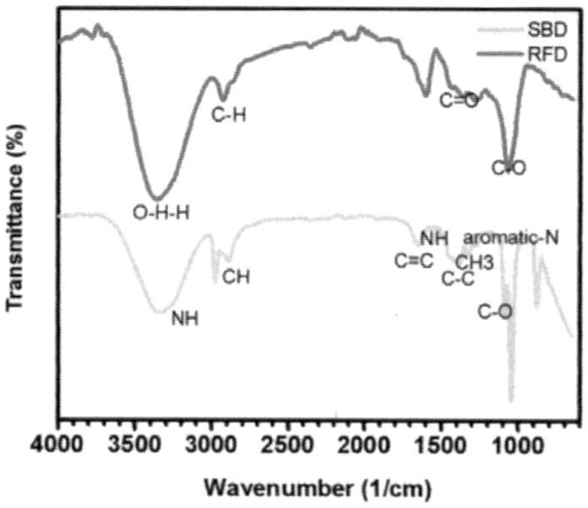

(b)

Figure 3. (a) UV-Vis absorption spectres and (b) Fourier Transforms Infra-Red spectres of RFD and SBD.

Maximum absorption peak of SBD was 480.5nm. Chlorophyll absorbs blue and red light strongly and transmits green light which make them looked green in colour, while RFD which was anthocyanin pigment absorbs green light. The band gap of RFD and SBD were determined using eq (1) [15]. Where E_g is band gap, h is plank constant, while c is speed of light and λ_max is wavelength of maximum peak of UV-Vis absorption spectra.

$$E_g = \frac{hc}{\lambda_{max}} \quad (1)$$

The calculated band gap of RFD was 2.29eV while SBD was 2.58eV. The collected Fourier Transforms Infra-Red (FTIR) spectra of RFD and SBD are shown in Figure 3(b).

The FTIR spectrum of RFD exhibits most of the characteristic of the anthocyanin. Peak 3334 cm-1 represented aromatic O-H-H while peak 2962 cm-1 corresponded to C-H stretching. Other peaks like 1639cm-1 was corresponding to characteristic of C=O stretching vibration of anthocyanin and peak 1053cm-1 represented C-O stretching. Lastly, the peak at 557cm-1 corresponded to C-Cl bond in anthocyanin [10]. The FTIR spectrum of SBD exhibits most of characteristic of chlorophyll. Peak 3346.5 cm-1 represented NH stretch and peak 2974.2 cm-1 corresponded to CH stretch. Peak 1651.1 cm-1 represented C=C while peak 1535.3 cm-1 corresponded to NH out of plane. Other peaks were 1415.8 cm-1, 1384.9 cm-1, 1327 cm-1 and 1085.9 cm-1 corresponded to aromatic C-C stretch, CH3, aromatic-N and C-O stretch respectively [16].

3.2. Cyclic-voltammetry of RFD and SBD

The Highest Occupied Molecular Orbital (HOMO) and Lowest Unoccupied Molecular Orbital (LUMO) were evaluated from oxidation onset and reduction onset of cyclic-voltammetry (C-V) plot in Figure 4. The characterization was executed with Platinum. Pt mesh and potassium chloride (KCl) as counter and references electrodes, respectively. The electrochemical measurement was carried out with the scan rate of 0.05V/s and step potential of 0.01V. All measurement was done at room temperature.

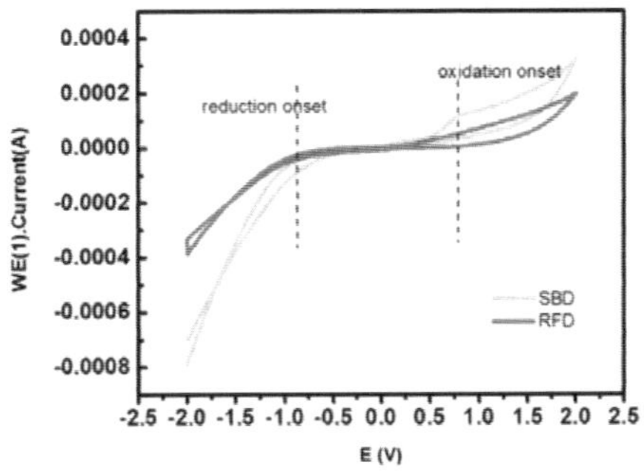

Figure 4. Cyclic-voltammetry plot of RFD and SBD.

The HOMO and LUMO of RFD and BSD were calculated using eq(2) and (3), where E_HOMO and E_LUMO represent energy level of HOMO and LUMO respectively, while E_Oxidation^Onset and E_Reduction^Onset are the value of oxidation and reduction coordinates in C-V plot respectively and 4.4 is adjustment factor [17-18].

$$E_{HOMO} = e[E_{Oxidation}^{Onset} + 4.4] \quad (2)$$

$$E_{LUMO} = e[E_{Reduction}^{Onset} + 4.4] \quad (3)$$

The obtained LUMO value for RFD was -3.54eV while SBD was -3.50eV. The HOMO value for RFD was -5.83eV and for SBD was -5.96eV. Band gap value of RFD and SBD was calculated using eq (4) [20].

$$E_g = E_{LUMO} - E_{HOMO} \quad (4)$$

The band gap values that were obtained through C-V characterization for RFD was 2.30eV while for SBD was 2.5eV. These band gap values were congruent with one that was obtained from UV-Vis spectroscopy characterization, hence, proved the calculated band gap of RFD and SBD were valid. From the calculation of HOMO and LUMO of RFD and SBD, the energy level alignment of every materials in proposed ZnO and RFD/SBD based DSSC can be elucidated.

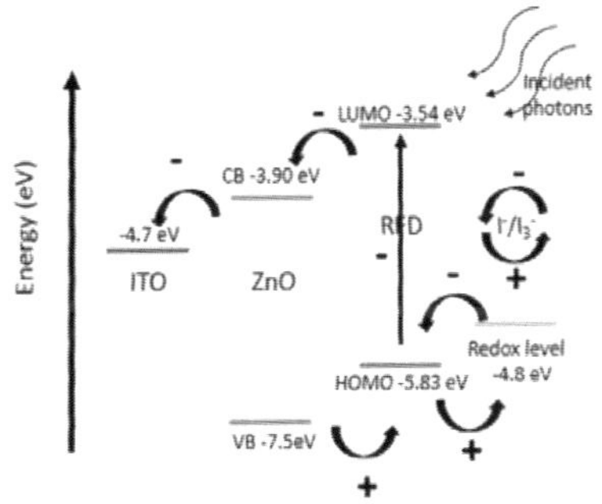

(a)

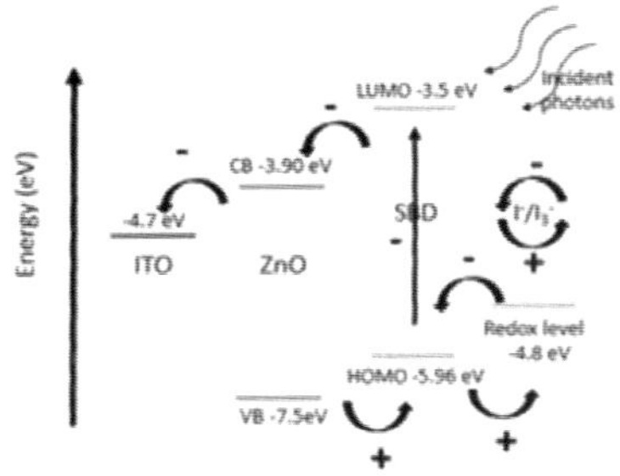

(b)

Figure 5. Energy level diagram and mechanism of photocurrent generation in ZnO based DSSCs with (a) RFD and (b) SBD.

IOP Conf. Series: Journal of Physics: Conf. Series **1027** (2018) 012016 doi:10.1088/1742-6596/1027/1/012016

By referring to Figure 5 the four steps of carrier movement in this DSSC can be assumed that upon incident photons from light illumination on DSSC, firstly the RFD/SBD molecule will gain enough energy to photoexcite its electron from HOMO to LUMO. Secondly, at LUMO the electron injects into the conduction band of ZnO and thirdly, diffuses into ITO conductive layer of glass substrate. After that, the electron will transfer back into the counter electrode (cathode) and diffused into the layer of catalyst metal then into the iodide electrolyte. The iodide electron will have redox reaction and donates electron to the oxidized RFD/SBD. The molecule of RFD/SBD will be back to its original state and theoretically, all of this mechanism will be repeated and sustained [19-21].

3.3. Dye-sensitized solar cell performances

DSSC samples that have been tested through I-V characterization exhibited the photodiode properties. The parameters of interest are open circuit voltage (V_{oc}), short circuit current (I_{sc}), maximum voltage (V_{max}), maximum current (I_{max}), fill factor (FF) and PCE as displayed in Table1. The FF and PCE were calculated using eq (5) and (6)

$$FF = \frac{I_{sc}V_{oc}}{I_{max}V_{max}} \tag{5}$$

$$\mathrm{PCE} = \frac{I_{max}V_{max}}{A_{active}P_{in}} \times 100 \tag{6}$$

where A_{active} is active area that has been illuminated with light source which is 2.8274 x10^{-5} m^2 and P_{in} is the intensity of light which is 100 W/m^2. I-V curves of DSSCs that have been sensitized with RFD in different duration is shown in Figure 6(a). The performance of RFD DSSCs increased up to sensitization time of 4h, then, gradually decreased when the sensitization duration increases to 6h and 8h. Therefore, the best sensitization duration in RFD was 4h. ZnO is sensitive to acidic media. RFD is anthocyanin that naturally acidic in nature (pH of RFD was 5.5) and it was suspected that Zn(II) complexes was formed after 4h that reduced electron injections and increased carrier combination that dampened the DSSC performance [21].

Figure 6(b) shows I-V curves of DSSCs that have been sensitized with SBD in different duration. The best I-V performance was from the sensitization of DSSC with SBD in 6h dipping duration. It was believed that, after 6h the SBD aggregated on ZnO nanoparticles and formed blocking layer for electron injection and mobility of it in DSSC and the most ideal of sensitization process is only monolayer of dye molecules required to anchor to metal oxide [22].

The changes pattern in I_{sc}, V_{oc}, FF and PCE of DSSC through different sensitization time by RFD and SBD can be understood more by referring to Table 1. DSSCs that have been sensitized with SBD had better performance than that of have been sensitized with RFD overall. It was assumed that since SBD is chlorophyll which absorb two range of visible light (red and blue regions), it harvested more photons from light compared to RFD which is anthocyanin that only absorb green light. Furthermore, chlorophyll is less acidic than anthocyanin (pH of SBD was 5.78) and less prone to form Zn(II) complexes that dampened the electron injection which will reduce the PCE value of DSSCs and allowed ZnO to be sensitized in longer period.

Table 1. I-V parameters of the fabricated DSSCs sensitized by RFD and SBD at different duration.

Dye	Sensitization duration (h)	V_{oc} (mV)	I_{sc} (μA)	V_{max} (mV)	I_{max} (μA)	FF	PCE (%)
RFD	2	205	1.06	100	0.57	0.38	0.002
	4	250	1.46	140	0.78	0.33	0.004
	6	105	0.87	55	0.44	0.34	0.001
	8	35	0.50	30	0.36	0.16	-
SBD	2	115	1.51	65	0.65	0.41	0.002
	4	245	1.43	135	0.75	0.35	0.004
	6	250	4.43	130	1.61	0.53	0.007
	8	75	0.54	45	0.30	0.30	0.001

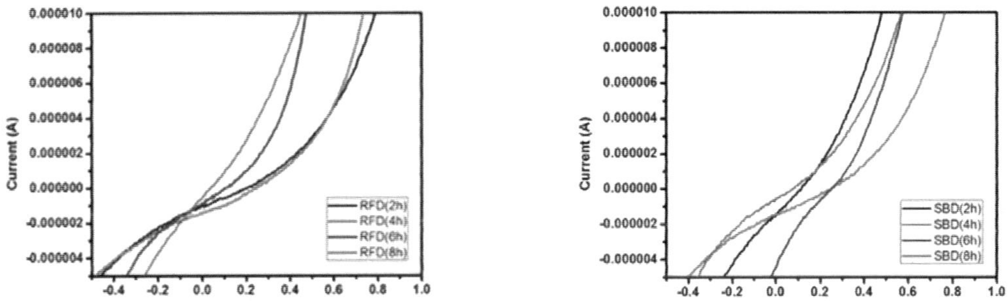

(a) (b)

Figure 6. Figure 6: I-V curves of DSSCs that have been sensitized with (a) RFD and (b) SBD in different duration.

4. Conclusion

In summary, ZnO and RFD/SBD based DSSCs were successfully fabricated. ZnO acted as electron collector active material while RFD/SBD served as electron donator active material in photoanode of DSSCs. RFD and SBD are two different plant pigments which are anthocyanin and chlorophyll respectively that harvested light at different region in visible range. SBD-DSSCs had better performance than RFD-DSSC due to the ability of chlorophyll to harvest red and blue light compared to that of RFD which is anthocyanin that only can harvest green light only. Another reason of better performance in SBD-DSSCs was due to less acidic nature of chlorophyll compared to anthocyanin, which less formation of Zn(II) complexes that hindered better performance of DSSCs.

Acknowledgement

The authors would like to acknowledge MyPhD scholarship for financial support.

References

[1] O'Regan B and Gratzel M 1991 *Letters to Nature* **353** 737
[2] Senthil T S, Muthukumarasamy N, Velauthapilhlai D, Agilan S, Thambidurai M and Balasundaraprabhu R 2011 *Renewable Energy* **36**(9) 2484–2488
[3] Zhou H, Wu L, Gao Y and Ma T 2011 *Journal of Photochemistry and Photobiology A: Chemistry* **219**(2–3) 188–194
[4] Hug H, Bader M, Mair P and Glatzel T 2014 *Applied Energy* **115** 216–225
[5] Sundara Venkatesh P, Balakumar S and Jeganathan K 2014 *RSC Advances* 4(10) 5030
[6] Zhang Q, Dandeneau C S, Zhou X and Cao G 2009 *Adv Mater (Weinheim, Ger.)* **21**(41) 4087–4108
[7] Tributsch H and Calvin M 1971 *Photochemistry and Photobiology* **14**(2) 95–112

IOP Conf. Series: Journal of Physics: Conf. Series **1027** (2018) 012016 doi:10.1088/1742-6596/1027/1/012016

[8] Dhafina W A, Ghapur E A and Hasiah S 2012 ARPN *Journal of Science and Technology* **2**(5) 432–436
[9] Son D Y, Bae K H, Kim H S and Park N G 2015 *The Journal of Physical Chemistry C* **119**(19) 10321–10328
[10] Dumbravă A, Georgescu A, Damache G, Badea C, Enache I, Oprea C and Gîrţu M A 2009 *Journal Of Optoelectronics And Advanced Materials* **10**(11) 2996 – 3002
[11] https://chem.libretexts.org/Textbook_Maps/General Chemistry Textbook Maps/Map%3A Chemistry Atoms First (OpenSTAX)/19%3A Transition Metals and Coordination_Chemistry
[12] Shanmugam V, Manoharan S, Anandan S and Murugan R 2013 *Spectrochimica Acta - Part A: Molecular and Biomolecular Spectroscopy* **104** 35–40
[13] Glover B J and Martin C 2012 *Current Biology* **22**(5) 147–150
[14] Lichtenthaler H K and Buschmann C 2005 *Handbook of Food Analytical Chemistry* 2–2 171–178
[15] Dharma J, Pisal A and Perkin Elmer I 2012 *Global Application Library* 4–7
[16] http://irpal.soft112.com/
[17] Arifin Z, Soeparman S, Widhiyanuriyawan D, Sutanto B and Suyitno 2017 *International Journal of Photoenergy* 30123
[18] Leonat L, Sbarcea G and Branzoi I V 2013 *U.P.B.Sci.Bull.,* Series B **75**(3) 111–118
[19] Dhafina W A, Hasiah S and Daud M Z 2016 *Chemical Reviews* **2**(4) 516–521
[20] Narayan M R 2012 *Renewable and Sustainable Energy Reviews* **16**(1) 208–215
[21] Anta J A, Guillén E and Tena-Zaera R 2012 *The Journal of Physical Chemistry C* **116**(21) 11413-11425
[22] Tagliaferro R, Colonna D, Brown T M, Reale A and Di Carlo A 2013 *Optics Express* **21**(3) 3235–42

IOP Conf. Series: Journal of Physics: Conf. Series **1027** (2018) 012017 doi:10.1088/1742-6596/1027/1/012017

Classification of alloys using laser induced breakdown spectroscopy with principle component analysis

Aneez Syuhada Mangsor[1], Zuhaib Haider Rizvi[1], Kashif Chaudhary[1,2], Muhammad Safwan Aziz*[1,2]

[1]Physics Department, Universiti Teknologi Malaysia, Johor Bahru, Malaysia.
[2]Laser Centre, Ibnu Sina Institute for Scientific and Industrial Research (ISI-SIR), Universiti Teknologi Malaysia (UTM), Johor Bahru, 81300, Malaysia.

Corresponding author: safwanaziz@utm.my

Abstract. The study of atomic spectroscopy has contributed to a wide range of scientific applications. In principle, laser induced breakdown spectroscopy (LIBS) method has been used to analyse various types of matter regardless of its physical state, either it is solid, liquid or gas because all elements emit light of characteristic frequencies when it is excited to sufficiently high energy. The aim of this work was to analyse the signature spectrums of each element contained in three different types of samples. Metal alloys of Aluminium, Titanium and Brass with the purities of 75%, 80%, 85%, 90% and 95% were used as the manipulated variable and their LIBS spectra were recorded. The characteristic emission lines of main elements were identified from the spectra as well as its corresponding contents. Principal component analysis (PCA) was carried out using the data from LIBS spectra. Three obvious clusters were observed in 3-dimensional PCA plot which corresponding to the different group of alloys. Findings from this study showed that LIBS technology with the help of principle component analysis could conduct the variety discrimination of alloys demonstrating the capability of LIBS-PCA method in field of spectro-analysis. Thus, LIBS-PCA method is believed to be an effective method for classifying alloys with different percentage of purifications, which was high-cost and time-consuming before.

1. Introduction

Laser Induced Breakdown Spectroscopy (LIBS), also known as laser induced plasma spectroscopy (LIPS), is an elemental analysis technique applicable for both qualitative and quantitative analysis of samples with unknown composition [1]. LIBS technique for the determination of elemental composition of materials in the form of solids, liquids and gases was first discovered on 1960's [2]. Researches in LIBS have reaching its peak on 1970 and continued to grow during the past decades due to the factor of it is a method in which numerous application have been realised. For instance, number of papers published on LIBS in the year 2005 was about 350 and in 2007, the number of papers related to fundamentals and applications of LIBS reached the number of 600 [3]. The development on LIBS during these recent years had narrowed down the gap performance of LIBS compared to the other well-known approaches used for the analytical atomic spectrometry of the samples [4]. The examples of the conventional methods are inductively coupled plasma-atomic emission spectrometry (ICP-AES), electro thermal atomization-atomic absorption spectrometry (ETA-AAS), and inductively coupled plasma-mass spectrometry (ICP-MS). It is said that LIBS is a more effective and up-to-date technique suitable for faster processing of data compared to the conventional methods. LIBS had been recognized as an attractive and promising technique to be used for the fast and whole chemical analyzation due to its "put & play" technique suitable for a wide range of applications [5].

Principle Component Analysis (PCA), also known as Karhunen-Loeve expansion is a technique where it is considered as a classical feature extraction and data representation. By using this technique, the variation can be emphasized and the strong patterns of the dataset can be visualized. PCA make it possible to represent the pattern of the similarity of the observations and it will be viewed in points on a map [6]. Multivariate analyses have been already applied for the classification purposes of LIBS samples [7] such as the classifications of samples using partial least squares discriminant analysis (PLS-DA) [8] and soft independent modeling of class analogy (SIMCA) [9]. However, few studies on variety classification of alloys using LIBS couple with principle component analysis (PCA) have been reported.

This research aimed to analysis the spectral lines of the target samples and to distinguish between metallic alloys using LIBS-PCA technique. In this study, the pattern of the spectrum lines of each target samples will be observed. Nd:YAG laser pulse (1064 nm, 740 mj, 6ns) will be used to produce plasma on different types of alloys which are Aluminium, Titanium and Brass with the metallic purities of 75%, 80%, 85%, 90% and 95%. This research is designed to contribute to the existing knowledge on spectral analysis of laser induced plasma spectroscopy. This research aimed to explore the characteristics of various spectral lines associated with atomic and ionic transitions of different types of alloys that are susceptible to variations in plasma conditions. It will help in the understanding of the dynamics of spectral line emissions from laser induced plasma and selection of lines to determine plasma parameters. The choice of metal alloys purities in this experiment is important to get the knowledge of the composition of the elements in the targeted sample based on different purities. This method is remarkably useful for the industries in the interest of removing the foreign or contaminating substances to get the optimum production of the output required.

2. Experimental

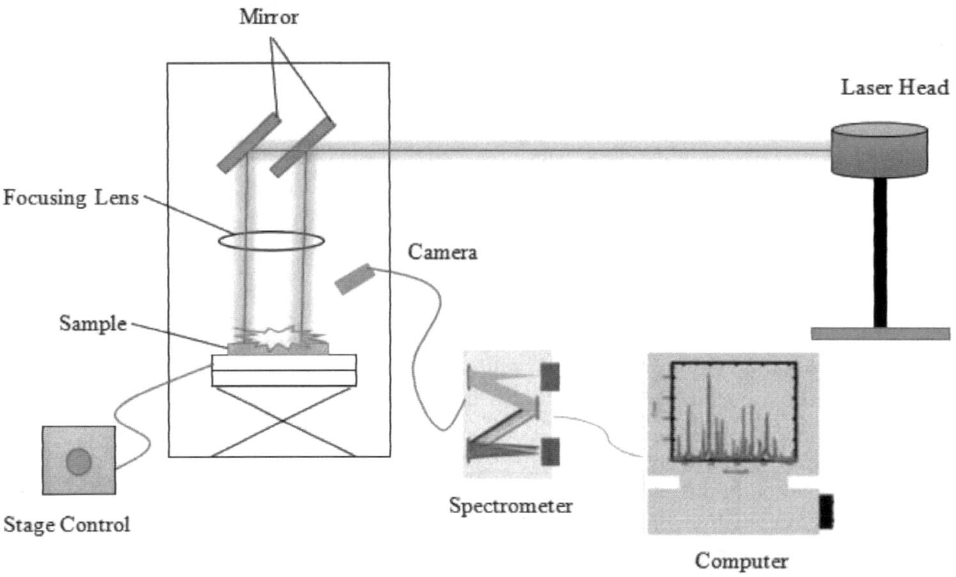

Figure 1. Schematic of LIBS experimental setup

Figure 1 shows the schematic diagram of LIBS experimental setup to generate laser induces plasma of alloy samples and capture the optical emissions using optical spectrometer. The experiments were conducted in air. Firstly, the focusing lens is cleaned using the propanol. This step is to make sure the lens is clean and free from dust. The apparatus setup is prepared as in figure 3.2.1. Nd:YAG laser (1064nm, 6ns), operating at fundamental harmonic is directed at 90° on alloy sample surface using plano-convex lens with focal length (10 cm). The lens position is optimized and readjusted to attain the best focusing. The target sample is placed on translation stage to expose the fresh position of target material and avoid crater. The environmental air is kept ambient. Radiations emitted from plasma are collected via collecting lens attached in front of a 600µm core diameter attached to Ocean Optics MayaPro spectrometer. Optical emission spectra rom plasma is collected in the direction parallel to the target surface.

For each and every alloy samples, the samples undergo at least five laser shot before the real shot. This step is to ensure dust and any contamination on the surface of the samples is totally penetrated. Only the elements contained inside the samples themselves are taken into account in the LIBS spectra. Each sample of the three alloys of different purity is exposed at one specific distinct spot on the surface to five laser shots. For each laser shot an emission spectrum from the plasma is recorded on the computer. Origin software was used to analyzed the spectra of laser induced plasma spectroscopy. Dataset with many variables will tend to measure the same driving principle governing the system. In the experiment, there are five different purities of 75%, 80%, 85%, 90% and 95% of three different alloys of Al, Brass and Cu. Even though the purities are different, the major element of each samples shows the overlapping content. This concept makes it relevant to apply the principle component analysis to group the dataset according to the family.

3. Result and discussion

Identification of lines is done by matching the spectral lines and relative intensities with the NIST atomic spectral database (https://www.nist.gov/pml/atomic-spectra-database) and data published in literature. The wavelength range for all LIBS spectra are starting at 200 nm in the ultraviolet (UV) and extending into the near-infrared (NIR) to 700 nm. Based on the graphs of all spectra, the high intensity spectral lines can be observed in the range in between 250nm and 600nm. The major peaks with high intensity value represent the major element contained in the samples. Several obvious peaks are observed that give indication of the presence of Aluminium, Brass and Copper even though the samples come from different purities.

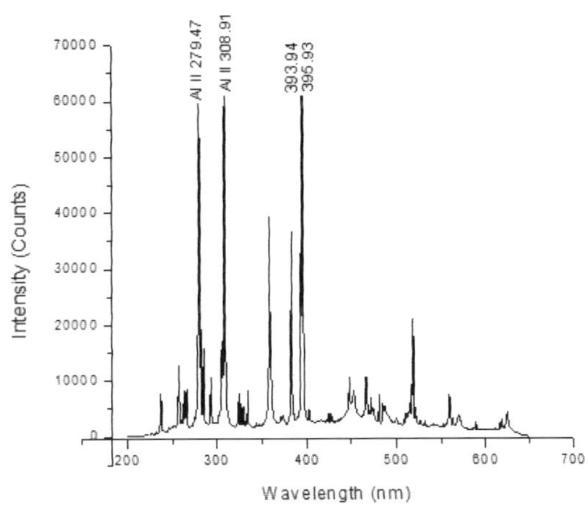

Figure 2. Optical emission of sample 1 Aluminium alloy

For Aluminium, the wavelengths of the four major peaks from five samples value are quite close to the value of Al II provided by the NIST which are 280.1178nm and 308.8516 nm. Figure 2 give an example of the optical emission of LIBS spectra of sample 1 of Aluminium with several obvious peaks around the wavelengths at 279.47nm, 308.91nm, 393.94nm and 395. 93nm.For Brass, the wavelengths of the five major peaks from five samples value are quite close to the value of Cu I provided by the NIST which is 521.8202nm. It is proven that the major peak gives meaning that the Cu is among the major element contained in the Brass samples. For Copper, the wavelengths of the three major peaks from five samples value are quite close to the value of Cu I provided by the NIST which is 521.8202nm.

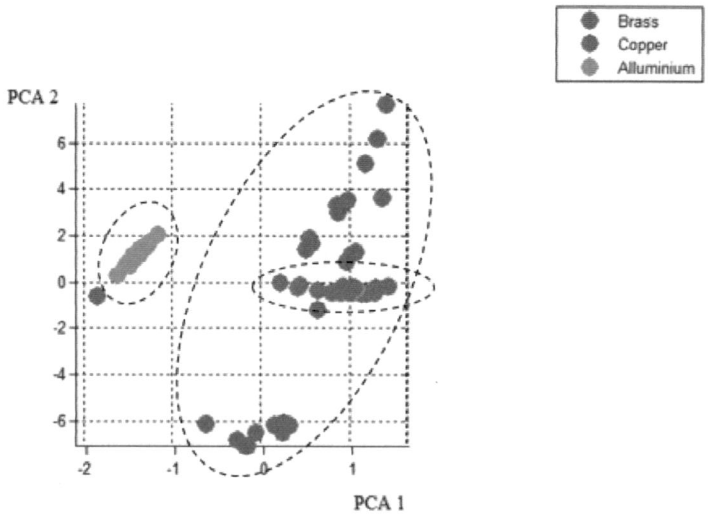

Figure 3. Principle Component Scatter Plot in two dimensional

75 samples in which 25 samples comes from each types of alloys are transformed into principle component (PCs). The principle component scatter plot was executed to display the variation among the three types of alloy samples. The 2 PCs explained the variations of the spectra as well as their scores. For one spectrum will produce one point on the score plot. Apparent clustering can be noticed from the figure 3 where there is an obvious difference between three groups of alloy samples. Al, Brass and Cu from different purifications are grouping into their family of major element. This is a proof that is relevant to show the effectiveness of PCA technique in discriminate group differences into their constituent major elements.

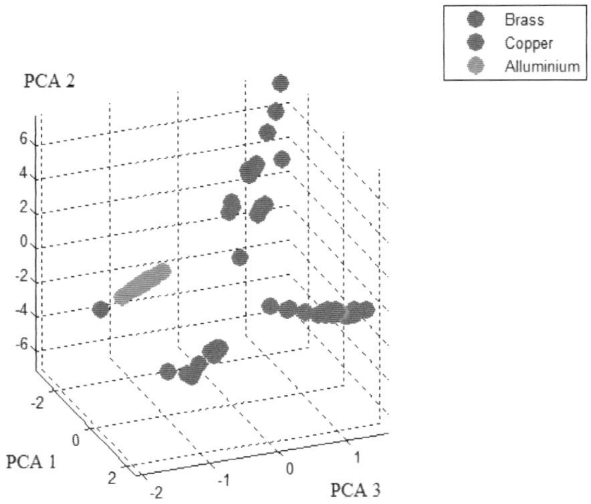

Figure 4. Principle Component Scatter Plot in three dimensional

PCA had compressed the variance of the LIBS spectra into three PCs showing the score plot of the three PCs in three dimensional view. It can be seen points of LIBS spectra were clustered in three group. It draws a clear boundary between the different types of samples.

4. Conclusion

Based on the LIBS spectra, the wavelengths of the major peaks have been identified. The wavelengths of the peak with high intensities are then being compared to the NIST data spectral database to identify the elements contained within the samples. PCA analysis then being carried out to the LIBS spectra. Three obvious clusters were observed and the clusters were analysed. Results demonstrated that LIBS-PCA is a combined technique applicable in distinguishing the samples according to the different variables contained within the studied samples. LIBS-PCA is a relevant technique to be used to identify the composition of the elements without consuming too much money and time.

References

[1] R J Lasheras, C Bello-Galvez, E M Roriguez-Celis and J Anzano 2011 Discrimination of organic solid materials by LIBS using methods of correlation and normalized coordinates. *Journal of hazardous materials* **192**(2) 704-713

[2] E R Runge, R W Minck and F R Bryan 1964 *Spectrochim. Acta. 20B 73*323.

[3] Noll R 2006 Terms and notations for laser-induced breakdown spectroscopy. *Analytical and bioanalytical chemistry* **385**(2) 214-218

[4] J D Winefordner, I B Gornushkin, T Correll, E Gibb, B W Smith and N Omenetto 2004 Comparing several atomic spectrometric methods to the super stars: special emphasis on laser induced breakdown spectrometry, LIBS, a future super star. *Journal of Analytical Atomic Spectrometry* **19**(9) 1061–1083

[5] B Kearton and Y Mattley 2008 Laser-induced breakdown spectroscopy: sparking new applications. *Nature Photonics* **2**(9) 537–540

[6] Jolliffe I T 2002 Principal Component Analysis. *New York: Springer*

[7] Martin M Z, Labbé N, Rials T G and Wullschleger S D 2005 Analysis of preservative-treated wood by multivariate analysis of laser-induced breakdown spectroscopy spectra. *Spectrochim. Acta B* **60** 1179–1185

[8] Zhang T, et. al. 201 Quantitative and classification analysis of slag samples by Laser-induced breakdown spectroscopy (LIBS) coupled with support vector machine (SVM) and partial least square (PLS) methods. *J. Anal. Atom. Spectrom* **30** 368–374

[9] Sirven J B, et. Al. 2007 Feasibility study of rock identification at the surface of Mars by remote laser-induced breakdown spectroscopy and three chemometric methods. *J. Anal. At. Spectrom* **22** 1471–1480